Welding Processes

Welding Processes

Editor

Thanh Phuc Van

scitus
academics

Welding Processes
Edited by **Thanh Phuc Van**

ISBN: 978-1-68117-663-5
Library of Congress Control Number: 2016946530

Reprint 2025

SCITUS Academics LLC,
616, Corporate Way, Suite 2,
4766, Valley Cottage, NY 10989

www.scitusacademics.com

Notice

Printed in United States of America on Acid Free Paper

Preface

Welding is a very diverse trade. Almost every manufactured product utilizes a welding or joining process in its production. Products that are produced by welding range from small objects, such as sunglasses and dental braces to larger objects. Over the years, a number of major innovations have been made in welding technology. Welding is a process of joining two metal pieces by the application of heat. Welding is the least expensive process and widely used now a days in fabrication. Welding joints different metals with the help of a number of processes in which heat is supplied either electrically or by mean of a gas torch. Different welding processes are used in the manufacturing of Auto mobiles bodies, structural work, tanks, and general machine repair work. In the industries, welding is used in refineries and pipe line fabrication. There are more than thirty different types of welding, from simple oxy-fuel (using a flame) welding to high-tech processes such as laser beam welding.

Welding Processes is intended to provide valued information to undergraduate and graduate students, welding engineers, and researchers with updates on recent achievements in welding.

Table of Contents

1

CHAPTER

Remote Laser Welding with High Power Fiber Lasers

Paul Kah, Jinhong Lu, Jukka Martikainen, Raimo Suoranta

Laboratory of Welding Technology and Laser Processing, Department of Mechanical Engineering, Lappeenranta University of Technology, Lappeenranta, Finland

ABSTRACT

The introduction of the high power fiber laser with brilliant beam quality has enabled the rapid development of remote laser welding (RLW). This paper presents a theoretical review of remote laser welding. As a promising technology, RLW offers increased flexibility, high operational speed, and reduced cycle time to process a wide range of workpieces. This study presents the typical characteristics of RLW with high power fiber lasers. It also investigates the influence of process parameters such as laser power, welding speed, shielding gas supply, beam inclination and focal position on the weld seam quality.

Keywords: Remote Laser Welding; Scanner; Fiber Laser Process Parameters

1. INTRODUCTION

Remote laser welding (RLW), a non-contact robotic laser welding technique, has been developed and implemented to improve the productivity and flexibility of conventional laser welding [1]. The first experimental analysis of RLW was reported by John Macken in 1996 [2]. The remote welding process is characterized by a long focal length (up to 1600 mm), a high-power and brilliant-quality laser source, and beam deflection by the scanner [3-5]. Compared to conventional laser welding, the remote welding technology offers increased flexibility, higher working speed, and reduced cycle time [3,6].

Thus far, RLW has been studied and implemented with a variety of high-power laser sources. In [1], the RLW systems are implemented with CO_2 and Nd: YAG lasers. In [7], the RLW experiments are carried out with Nd: YAG and disk lasers. RLW applications with highpower fiber lasers have been introduced in [6,8]. With a wavelength of 1080 nm, these high-power fiber lasers enable beam delivery through the fiber, enhancing spatial flexibility and accurate focusing. By contrast, it is impossible to deliver the beam of CO_2 lasers via an optical fiber due to its rather long wavelength of 10.6 μm [7]. The suitable wavelength of high-power fiber lasers, combined with excellent beam quality, makes them a promising alternative to conventional CO_2 and Nd: YAG lasers in RLW applications.

The operating principle of RLW is normally based on using a scanner to deflect and position the laser beam onto the surface of the workpiece travelling at high speed [9] and, at present, 2D-scanners are the most widely adopted scanners in remote welding applications. The 2D-scanner unit is a galvanometer system, in which two lightweight mirrors are used and rotated by motors. The system can handle a laser power of up to 5 kW and is more economical than 3D-scanners [10].

Remote laser welding the technology is not yet very widespread, although it clearly has potential in the automotive industry, such as in the seating, body in white and interior parts. RLW has replaced resistance spot welding with its increased laser usability rate and reduced process times in car body construction [6].

However, RLW still faces many challenges; in predicting process behavior, in ensuring acceptable and reliable weld quality, and in dealing with issues such as shielding gas supply, clamping and coated sheet metals [11,12]. Compared to conventional laser welding, a greater number of process parameters have to be taken into account in remote welding applications. These parameters can be primarily divided into the areas of beam quality, processing parameters, and material properties [6].

2. REMOTE LASER WELDING

2.1. Principle

Remote laser welding is not a new technology, but is based on the principle where a scanner deflects and positions the focused laser beam over the workpiece from a distance of typically 1000 - 1600 mm [13]. The first experiment in laser keyhole welding with 1600 mm focal length, by John Macken in 1996, is acknowledged as forming the cornerstone of remote laser welding technology [11]. The scanner enables the translation of the laser beam into large working areas of 1m × 1 m or over 2 m^3 3D working volume with a welding speed of up to 30 m/min [14].

Remote laser welding can be principally implemented with two modes; the scanner-integrated system and the robot-based system [15]. The scanner-integrated system utilizes a scanning unit (usually a 2D scanner) for positioning and focusing of the laser beam, as shown in Figure 1(a) [11]. RLW with the robot-based system is accomplished with a long focal length laser optic and a 6- axis robot, in which the robot serves for positioning of the laser beam on the surface of the workpiece, as illustrated in **Figure 1**(b) [11].

Compared to robot-based RLW, the scanner-integrated RLW system offers shorter processing times and higher accuracy in many applications. However, the laser beam quality requirements of scanner-integrated RLW are much higher than for robot-based RLW systems. **Table 1** presents the typical performance of scanner-integrated and robot-based RLW systems versus conventional laser welding.

2.2. Requirements

The typical requirements for RLW to make sound welds can be divided into three categories: the scanner for delivering and positioning the beam adequately, a highpower laser with sufficient quality for the long focal length of the system, and proper control of process parameters [16].

2.2.1. Scanner

The scanner is used to guide and rapidly position the beam on the surface of the workpiece along the desired weld path [16]. The lightweight and highly dynamic scan head enables extremely fast movement of the laser beam between welds, which means that positioning requires much less time than in the conventional laser welding process [17].

Figure 2 shows the typical elements of a scan head [9]. A scanner unit consists of a group of mirrors and lenses [9,18]. In remote welding, the laser beam first passes through the lenses. The smaller lens moves along the optical axis in order to change the focal position. Afterwards, the laser beam is deflected and guided successively by the mirrors X and Y. Finally, the laser beam is focused on the workpiece precisely along the desired weld seams [9]. **Figure 3** illustrates a classic 2D-scanner system including a seam tracking sensor, a line projector, and a high dynamic 2D-scanner unit [19].

2.2.2. High-Power Fiber Laser

High-power fiber lasers have recently been developed with attractive characteristics for materials processing applications [20]. High-power fiber lasers with brilliant beam quality can produce ultra-high peak power density of several MW/mm^2, which is important for high speed RLW with long focal length [21].

Table 1. Comparison of remote and conventional laser welding processes [4,13,15].

Typical performances	Conventional laser welding	Scanner-integrated RLW	Robot-based RLW
Cycle time	1	3	2
Accuracy	3	3	2
Accessibility	1	3	3
Flexibility	2	3	3
Programming effort	2	1	2
Invest	3	2	3
Maintenance	3	2	3

1 = poor/low, 2 = medium, 3 = good.

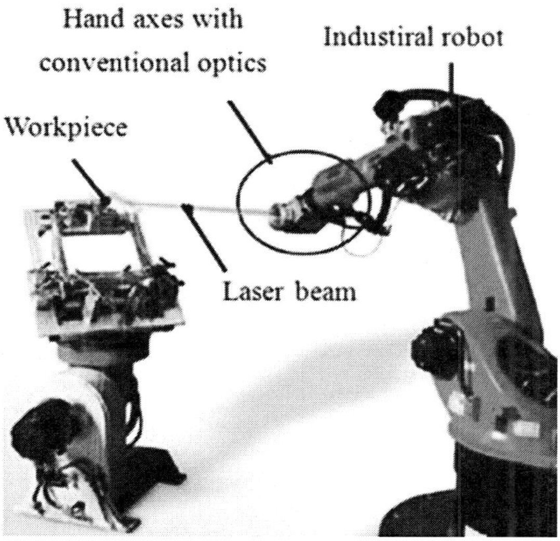

(a)

(b)

Figure 1. (a) Scanner-integrated RLW system and (b) Robot-based RLW system without scanner optics [11].

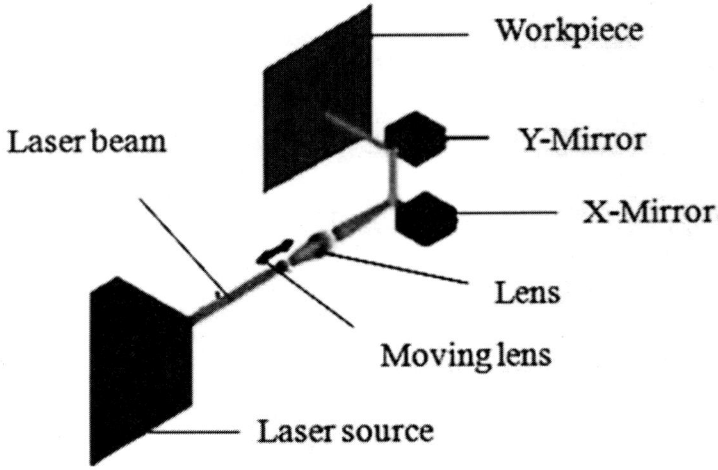

Figure 2. Typical elements of a scan head [9].

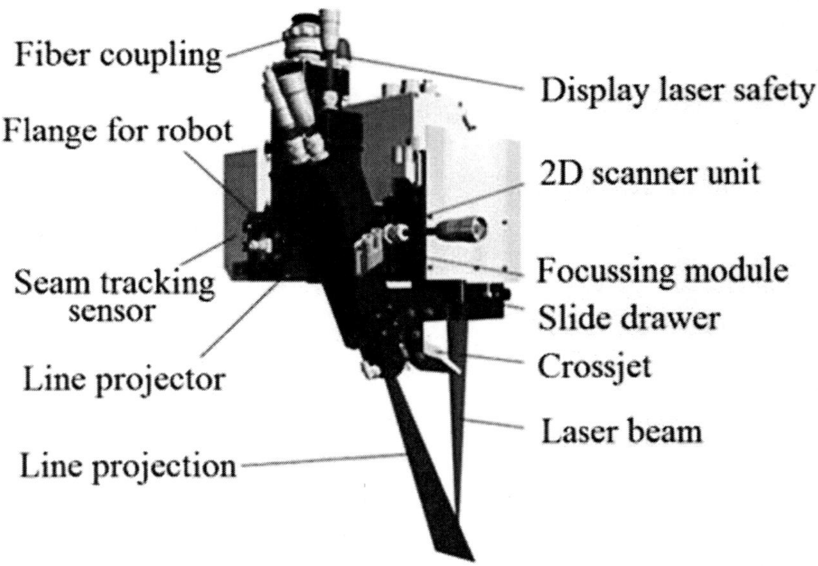

Figure 3. Main components of the 2D-scanner system [19].

Based on a number of studies, high-power fiber lasers have multiple advantages, such as [22,23]:

- High electric efficiency.
- Excellent beam quality.
- Relatively low operational costs due to the long lifetime.
- High flexibility in production because of the beam delivery with fibers.
- High absorption coefficient for thin sheets of most metals.
- Compact design and mobility.

Figure 4 shows an RLW cell with a high-power fiber laser anchored by a robot and equipped with welding head fixtures. It has been shown that remote fiber laser welding allows an increase in weld processing speeds, a reduction in consumables, such as weld wire and weld guns, less tooling and part fixturing, and a decrease in on-going maintenance costs typically associated with conventional welding processes [24].

2.3. Typical Characteristics of RLW

Remote laser welding integrated with a scanner offers a number of advantages over conventional laser welding, for example, high flexibility, reduced cycle time, high speed production of sound welds with consistent quality, a high degree of automation, good accessibility to the weld joints due to the long focal length, cost savings through the use of less materials and reduced maintenance demands; low distortion from a smaller heat input, and small floor space requirements [25,26].

2.3.1. High Productivity of RLW

Scanner-integrated RLW offers high productivity on account of its high welding velocity and reduced cycle time. The welding speed is dependent on the laser power used, which is normally 6 m/min and up to 30 m/min for special plates [27]. A remote fiber laser welding system can weld more than twice as fast as either MIG or resistance spot welding [24]. The elimination with the scanner of non-productive time means that the cycle time of RLW can be reduced by up to 80% [4,27]. The continuous motion of the scanner head in combination with the rapid positioning makes the RLW system move the laser beam between welds in less than 50 milliseconds, while robot positioning typically takes 2 - 3 seconds [3,17]. Hence, the cycle time of the scanner-integrated RLW is 6 - 10 times faster than resistance or arc welding processes [3].

Figure 4. RLW cell with a 5 KW IPG fiber laser [24].

2.3.2. High Flexibility of RLW

The long working distance between the scanner head and the workpiece in RLW enables high flexibility in the processing of joints that conventional laser welding cannot reach. As illustrated in**Figure 5**, the seam can only be welded with a long working distance. Another benefit of the long working distance is the avoidance of plume and spatters from optical elements [3,11]. The long working distance also allows the use of simpler clamping fixtures and makes RLW more profitable in big-size product applications [15].

ˀ.3.3. Low Costs of RLW

˙ has shown great potential for cost savings in the body shop. Based on ʰer of studies [3,12,25,28], the RLW system can decrease ˑring costs in many ways: reduction of average operating costs by ˑtion, reduction of costs by reduced number and increased efficiency of the laser guns, reduction of materials usage by a narrower joint overlap, and low maintenance costs due to the long lifetime.

Nevertheless, RLW also has disadvantages, for example, in regard to shielding gas supply, laser quality demand, clamping, and seam tracking [15,29]. Shielding gas supply is the biggest challenge for RLW, primarily because of the long working distance. The shielding gas has to be supplied to the workpieces independently [4,30]. The assistant gas does not move along with the welding seams but keeps the entire processing zone in a shielded environment. A number of nozzles are commonly required to work together simultaneously with a high gas flow rate, in order to protect the welding zone from oxidation and remove contamination. Therefore, the shielding gas consumption in RLW is high, which adversely affects total operational costs [11,12].

2.4. Applications of RLW

Currently, RLW is mainly applied in the mass production of flat assembly groups in which a high number of welding seams are being processed [11]. One typical area benefiting from RLW is the automotive industry [13,25]. Today, nearly 70 RLW systems are used for many automotive applications, such as seating (recliners, frames, tracks, panels), body in white (trunks, rear panels, doors/hang on parts, side walls, pillars, heat exchangers), and interior fittings (IP beams, rear shelf/hat rack) [16, 31].

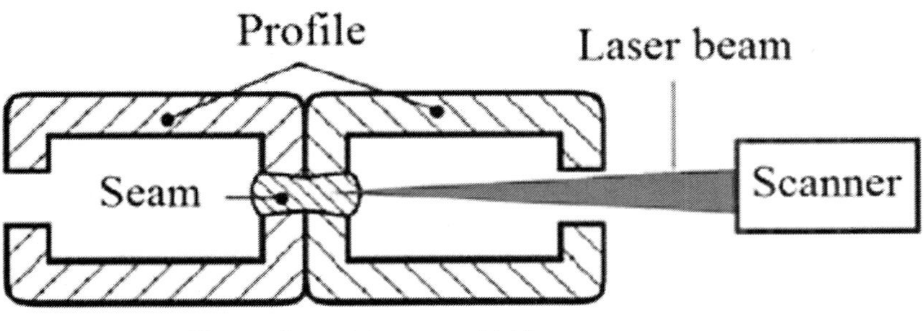

Figure 5. RLW of limited working zone [11].

3. PROCESS PARAMETERS OF RLW

Process parameters such as speed, focal position, beam inclination, shielding gas, and joint design, as well as material properties have been examined and studied in efforts to improve beam characteristics such as laser power, beam quality, wavelength, focal diameter and focal length, mode and polarization [32-34]. The key parameters of RLW, that is, boundary conditions, process conditions and process quality, as illustrated in **Figure 6**, have been researched and reported by Oefele et al. [6].

RLW with fiber lasers has been studied with four types of steels: mild steel [6,15], galvanized steel [35], galvannealed steel [8], and stainless steel [36]. The main processing parameters are given in**Table 2**.

Table 2. Process parameters of RLW with fiber lasers.

No.	Material [thickness in mm]	Power [kW]	Focal length [mm]	Focal diameter [μm]	Speed [m/min]	Focal position [mm]	Year [Reference]
1	Mild steel S355 [1.5 mm]	(MM) 8	470	640	2 - 14	0	2008 [6]
2	Mild steel S235	(MM) 8	1053	300 - 800	2 - 22	−80 - 80	2007 [15]
3	Galvanized steel JAC270EN [1.5 mm]	4	1250	360	5	0	2009 [35]
4	Galvannealed steel GA60 [1 mm]	4	1500	-	4	0	2009 [8]
5	Stainless steel 1.4301 [20 mm]	(SM) 3	500	109	4–10	0	2009 [36]

A number of processing parameters of RLW with fiber lasers have been investigated, such as laser power [6], welding speed [36], shielding gas supply [15], beam inclination [6], and focal position [15].

3.1. Laser Power

In laser processing, increased power of the laser gives increased welding speed. It has been reported that an increase in fiber laser power from 4 kW to 8 kW for the remote welding of mild steels with a constant focal diameter will result in increased welding speed. On the other hand, an increase in spatter, seam sagging, and decrease in seam quality will be associated with the increase of welding power [6,37].

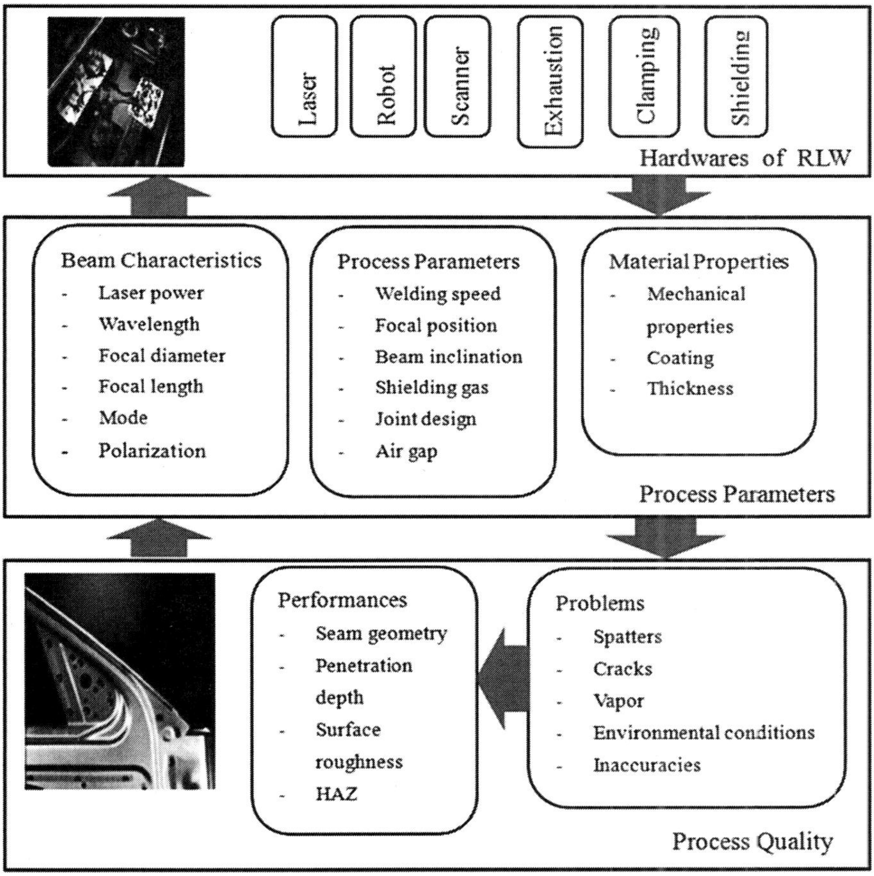

Figure 6. Key parameters of RLW [6].

Figure 7 illustrates the influence of laser power on welding speed and penetration in the remote fiber laser welding of mild steel S355. It can be seen that an increase in laser power results in a higher welding speed at the same penetration depth [6].

3.2. Welding Speed

Welding speed is one of the most important factors in RLW. It is generally dependent on the power density, required penetration, laser mode, focal position, and the number of rescans. Welding speed has an effect on the weld pool, weld bead shape, and penetration depth [33].

Figure 8 shows the influence of welding speed on the welding depth in the remote fiber laser welding of stainless steel 1.4301 with different laser powers. It can be seen that the welding depth decreases with increase in welding speed. At the same time, the welding depth increases with increases in laser power [36].

3.3. Shielding Gas Supply

Shielding gas supply offers numerous benefits to RLW: protection of the interaction zone from oxidation, removal of metal vapor and particles from the laser beam path, and stabilization of the welding process from plasma initiation [38]. The most common shielding gases used in RLW of steels are argon, helium, nitrogen, and compressed air [39]. A typical shielding gas flow in RLW is approximately 15 - 20 l/min, depending on the length of the welds [4].

The influence of shielding gas on the welding depth and speed of high power remote fiber laser welding is shown in **Figure 9**. The penetration depth decreases constantly with increase in the welding speed, with or without shielding gas. The influence of shielding gas on the penetration depth is very strong when the welding speed is less than 5 m/min. The main reason is probably the great plume of metal vapor produced in the keyhole, which hinders the welding process. With a welding speed over 14 m/min, the shielding gas has an insignificant effect on the penetration depth of RLW [15].

3.4. Beam Inclination

RLW enables a wide range of beam inclination angles, due to the long working distance. Depending on the inclination angle, the laser beam has different spot areas and shapes on the surface of the workpiece. Beam inclination has an effect on the welding depth and interface width through varied power densities and absorption. An inclination angle of more than 60° has a strong influence on the penetration depth. The inclination range of 20° - 30° has a negligible effect on the welding process. It is reported that inclination angles of 0° - 20° have no influence on weld quality [6].

3.5. Focal Position

The focal position has an effect on the penetration depth, bead shape, and the quality of cut kerf. The focal positions of the laser beam vary with different material thicknesses [40]. The recommended focal position to obtain maximum penetration is normally located in the workpiece, around 1 mm from the top surface [33].

Figure 10 shows the effect of focal position on the penetration in RLW of steel with disk lasers. It can be seen that 0 mm focal position from the top surface of the workpiece makes the deepest penetration. Welding penetrations decrease with increase in distance from the position at which the laser beam is focused [15].

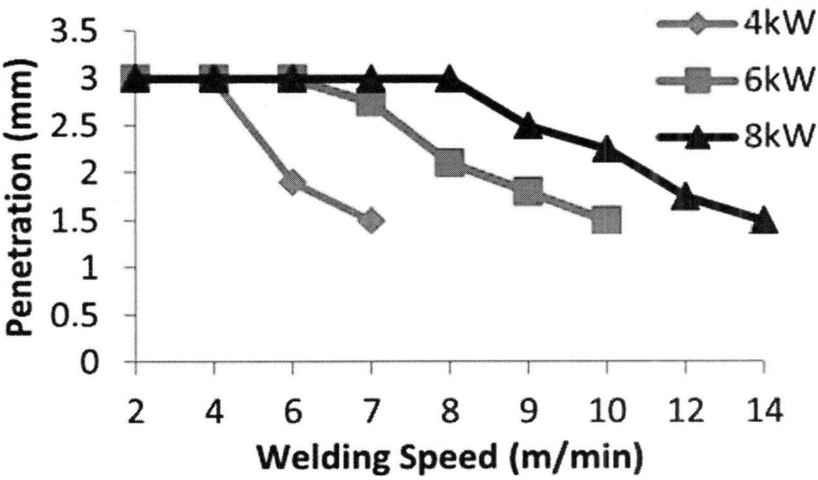

Figure 7. Influence of welding power on RLW; with exhaustion, no shielding gas, focal point is 640 μm; focal length is 470 mm; S355 mild steel [6].

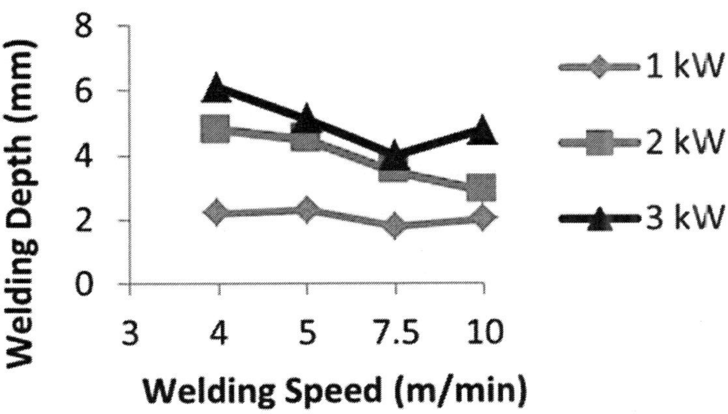

Figure 8. Influence of welding speed on the welding depth of RLW of stainless steel 1.4301 with different laser powers, the focal length is 500 mm [36].

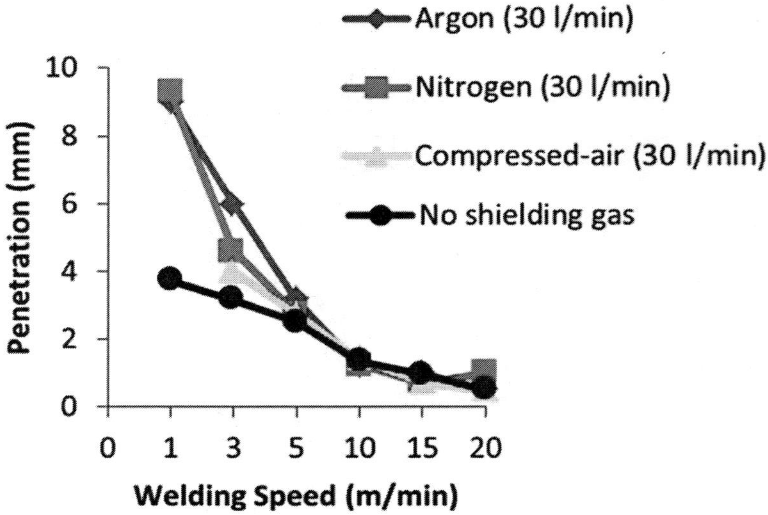

Figure 9. Influence of shielding gas on welding depth and welding speed of RLW of steel with an 8-kW fiber laser [15].

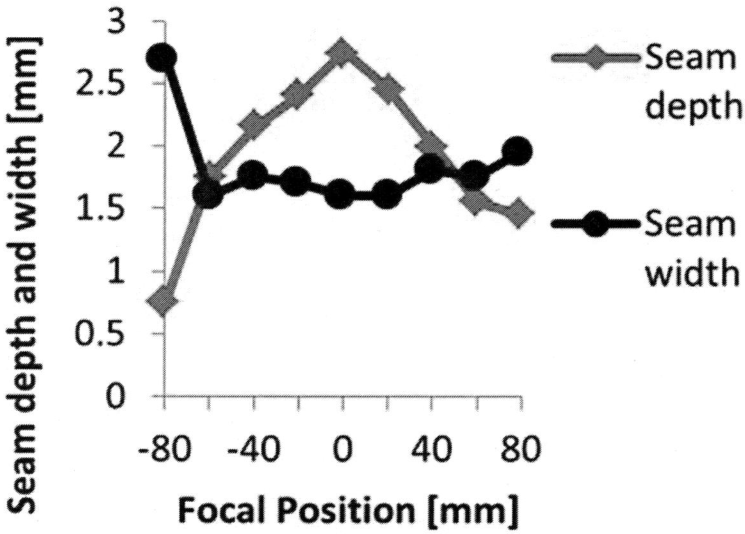

Figure 10. Influence of focal positions on the seam depth and width of remote fiber laser welding [15].

4. CONCLUSION

This paper studied remote laser welding with fiber lasers. Remote fiber laser welding is a versatile technology with a promising future. The process has been adopted by the automotive industry due to its advantageous characteristics, such as high productivity, high flexibility, and low operational costs. The scanner system is one of the most important requirements for remote fiber laser welding. However, to make a sound weld, a number of process parameters, such as beam quality, laser power, welding speed, shielding gas supply, and focal position have to be taken into account in remote laser welding.

REFERENCES

1. G. Tsoukantas, A. Stournaras and G. Chryssolouris, "Experimental Investigation of Remote Laser Welding with CO_2 and Nd: YAG Laser-Based Systems," Journal of Laser Applications, Vol. 20, No. 1, 2008, pp. 50-58.
2. J. Macken, "Remote Laser Welding," Proceeding of the International Body Engineering Conference, Advanced Technologies and Processes, Warren-MI, 1996, pp. 11- 15.
3. J. Cann, "A Look at the Remote Laser Beam Welding," Welding Journal, Vol. 84, No. 8, 2005, pp. 34-37.
4. M. Grupp, T. Seefeld and F. Vollertsen, "Laser Beam Welding with Scanner," Proceedings of the Second International WLT-Conference on Lasers in Manufacturing, Munich, 2003, pp. 211-222.
5. G. Tsoukantas and G. Chryssolouris, "Theoretical and Experimental Analysis of the Remote Welding Process on Thin, Lap-Joined AISI 304 Sheets," International Journal of Advanced Manufacturing Technology, Vol. 35, No. 9- 10, 2008, pp. 880-894.
6. F. Oefele, J. Musiol and M. F. Zaeh, "Influence of Remote-Laser-Welding Parameters for an 8 kW Fibre Laser on Seam Quality of Steels," The 27th International Congress on Applications of Lasers & Electro-Optic (ICALEO 2008), Temecula, 2008, pp. 399-405.
7. C. Kim, J. Kim, H. Lim and J. Kim, "Investigation of Laser Remote Welding Using Disc Laser," Journal of Materials Processing Technology, Vo. 201, No. 1-3, 2008, pp. 521- 525.

8. S. Kenji, Y. Motomichi, M. Kenta, Y. Daisuke, K. Yusuke and O. Takuya, "Development of Remote Laser Welding Method Using Long Focal-Distance Lens for Automobile Galvannealed Steel," Transactions of the Japan Welding Society, Vol. 27, No. 2, 2009, pp. 60-63.

9. K. Klingbeil, "What You Need to Know about Remote Laser Welding: A Look at How Remote Laser Welding Works and How It Can Be Applied to Your Manufacturing Process," Welding Journal, Vol. 85, No. 8, 2006, pp. 44-46.

10. M. P. Vänskä, "Implementing the Modern Fiber Laser Technology for Welding of Stainless Tubular Products," Master's Thesis, Lappeenranta University of Technology, Lappeenranta, 2009.

11. M. F. Zaeh, J. Moesl, J. Musiol and F. Oefele, "Material Processing with Remote Technology—Revolution or Evolution?" Physics Procedia, Vol. 5, 2010, pp. 19-33.

12. T. Heston, "Remote Laser Beam Welding Shows Potential in the Body Shop," Welding Journal, Vol. 79, No. 6, 2000, pp. 39-42.

13. M. Bemenek, "Technology Report: Welding from a Distance," Industrial Laser Solutions, Vol. 21, No. 3, 2006, pp. 19-23.

14. E. Beyer, A. Klotzbach, V. Fleischer and L. Morgenthal, "Nd: YAG-Remote Welding with Robots," Proceedings of the Second International WLT-Conference on Lasers in Manufacturing, Munich, 2003, pp. 367-373.

15. M. F. Zaeh, U. Munzert and F. Oefele, "Robot Based Remote-Laser-Welding without Scanner Optics," Proceedings of the Fourth International WLT-Conference on Lasers in Manufacturing, Munich, 2007, pp. 1-8.

16. D. A. Sabo, "The Evolution of Scanners for Remote Welding Applications: The Rise of Beam Quality Leads to Proliferation of Remote Welding Applications," 2007. http://www.thefabricator.com/article/lasercutting/the-evolution-of-scanners-for-remote-welding-applications

17. M. Bea, R. Brockmann and D. Havrilla, "Remote Laser Welding in Automotive Production," Industrial Laser Solutions, Vol. 26, No. 5, 2011, pp. 8-12.

18. M. F. Zaeh, J. Musiol and J. Moesl, "Methodical Qualification of Scanner Systems for Remote Laser Cutting," Proceedings of the 29th International Congress on Applications of Laser & Electro-Optics, Orlando, 26-30 September 2010, pp. 362-370.

19. Scansonic, "Remote Laser Welding—Adaptive," 2011. http://www.scansonic.de/files/downloads/scansonic_rlw-a_datasheet_v2.0_en.pdf

20. L. Quintino, A. Costa, R. Miranda, D. Yapp, V. Kumar and C. J. Kong, "Welding with High Power Fiber Laser— A Preliminary Study," Materials and Design, Vol. 28, No. 4, 2007, pp. 1231-1237.

21. Y. Kawahito, M. Mizutani and S. Katayama, "Investigation of High-Power Fiber Laser Welding Phenomena of Stainless Steel," Transactions of JWRI, Vol. 36, No. 2, 2007, pp. 11-15.

22. A. Kratky, D. Schuöcker and G. Liedl, "Processing with kW Fiber Lasers—Advantages and Limits," XVII International Symposium on Gas Flow, Chemical Lasers and High-Power Lasers, 21 April 2009.

23. M. Lütke, A. Mahrle, T. Himmer, L. Morgenthal and E. Beyer, "Remote Cutting—A Smart Solution Using the Advantages of High Brightness Lasers," Proceedings of the International Congress on Applications of Lasers & Electro-Optics, Temecula, 20-23 October 2008, pp. 695- 702.

24. Duggan Manufacturing, "New Process Development Facility for Remote Fiber Laser Welding," 2010. http://weldingdesign.com/equipment-automation/news/process-development-facility-remote-fiber-laser-0313/

25. P. Anthony, "The Reality of Remote Laser Welding," Industrial Laser Solutions, Vol. 19, No. 2, 2004, pp. 9-11.

26. A. Korinth and J. Cann, "Remote Welding Impacts on Auto Manufacture," Industrial Laser Solutions, Vol. 20, No. 5, 2005, pp. 26-31.

27. J. Hatwig, G. Reinhart and M. F. Zaeh, "Automated Task Planning for Industrial Robots and Laser Scanners for Remote Laser Beam Welding and Cutting," Production Engineering: Research and Development, Vol. 4, No. 4, 2010, pp. 327-332.

28. K. Mori, T. Tarui, T. Hasegawa and N. Yoshikawa, "Remote Laser Welding Applications for Car Bodies," Welding International, Vol. 24, No. 10, 2010, pp. 758-763.

29. L. X. Zhang, L. Wu, H. M. Gao and G. J. Zhang, "Agent- Based Modeling and Control of Remote Robotic Welding System," Robotic Welding, Intelligence and Automation, Vol. 362, 2007, pp. 187-194.

30. K. Krastel, "Remote Laser Welding in Industrial Applications," 2006. http://www.photonics.com/Article.aspx?AID=26648

31. G. Verhaeghe, "Remote Laser Welding for Automotive Seat Production," Industrial Laser Solutions, Vol. 27, No. 3, 2012, pp. 6-11.

32. J. Duong, "Qualification of the Convection Mode Laser Beam Welding Process," Master's Thesis, Lappeenranta University of Technology, Lappeenranta, 2010.

33. W. M. Steen and J. Mazumder, "Laser Material Processing," 4th Edition, Springer, London, 2010.

34. R. L. O'Brien, "Laser Beam Welding," In: R. L. O'Brien, Ed., Welding Handbook—Welding Processes, 8th Edition, Vol. 2, American Welding Society, Miami, 1991, pp. 713-738.

35. S. Katayama, S. Oiwa, N. Matsumoto, M. Mizutani and Y. Kawahito, "Fundamentals of Fiber Laser Remote Welding and Deep Penetration Welding," European Quantum Electronics Conference on Lasers and Electro-Optics, Munich, 14- 19 June 2009, p. 1.

36. L. Hartwig, R. Ebert, S. Kloetzer, S. Weinhold, J. Drechsel, F. Peuckert, J. Schille and H. Exner, "Material Processing with a 3 kW Single Mode Fiber Laser," Journal of Laser Micro/Nanoengineering, Vol. 5, No. 2, 2010, pp. 128-133.

37. F. Oefele, U. Munzert, M. F. Zaeh, "Remote-Laser-Welding with an 8 kW Fibre Laser," Proceedings of the Laser Assisted Net Shape Engineering, Erlangen, 2007, pp. 183- 196.

38. M. J. Song, B. H. Jung, M. Y. Lee and J. Suh, "Laser Welding Automobile Bumpers: Remote Welding System with Shielding Gas," Industrial Laser Solutions, Vol. 23, No. 4, 2008, pp. 17-20.

39. C. Dawes, "Laser Welding: A Practical Guide," Abington Publishing, Cambridge, 1992.

40. C. Wandera, "Laser Cutting of Austenitic Stainless Steel with a High Quality Laser Beam," Master's Thesis (In Finnish), Lappeenranta University of Technology, Lappeenranta, 2006.

2

CHAPTER

Submicron Particles during Macro- and Micro-Weldings Procedures in Industrial Indoor Environments and Health Implications for Welding Operators

Pasquale Avino [1,*], Maurizio Manigrasso [1], Pietro Pandolfi [2], Cosimo Tornese [2], Diego Settimi [2] and Nicola Paolucci [2]

[1]*DIT, INAIL Research Area, via IV Novembre 144, Rome I-00187, Italy*

[2]*Rheinmetall Italia S.p.A., Rome I-00100, Italy*

ABSTRACT

One of the emerging risks in the engineering and electronic industries is the exposure of workers to ultrafine particles during (micro-)welding operations, *i.e.*, processes used for joining two metal parts heated locally, which constitute the base metal, with or without addition of another metal which is the filler metal, melted between the edges to be joined. The process is accompanied by formation of metallic fumes arising from the molten metal as well as by the emission of metal fumes of variable composition depending on the alloys welded and fused. The aim of this paper is to investigate the number, concentration and size distribution of submicron particles produced by (micro-)welding processes. Particle number size distribution is continuously measured during (micro-)welding operations by means of two instruments, *i.e.*, Fast Mobility Particle Sizer and Nanoparticle Surface Area Monitor. The temporal variation of the particle number size distribution across the peaks evidences the strong and fast-evolving contribution of nucleation mode particles: peak values are maintained for less than 10 s. The implication of such contribution on human health is linked to the high deposition efficiency of submicronic particles in the alveolar interstitial region of the human respiratory system, where gas exchange occurs.

Keywords: welding; micro-welding; submicron particles; indoor; occupational; FMPS; ultrafine particles; deposition; lobar; dose

1. INTRODUCTION

In the engineering industry, a large variety of manufactures from small metal objects up to large machines is produced. The production cycles of the engineering industries are generally very simple. They start from raw materials that may be metals in bars, rods or coils whereas the operations involve cutting machines, bending, drilling, pressing, grinding, turning operations, use of milling cutters and drills. Work centers vary for preparing the various parts to be welded or screwed or assembled.

Among the various risks associated with these processes (e.g., generic dust from grinding and/or sanding, polishing, calendering, nuisance dust in the case of epoxy coatings, respirable dust, crystalline silica, mists of mineral oils, solvent vapors) [1], fumes from welding pose a complex risk [2,3,4,5,6,7]. In these fumes, various components may be present both in gaseous phase and solid state (particles at different size, ultrafine particles—nanoparticles, inhalable dust, respirable dust) [8,9].

The welding, one of the most common procedures in such industries [10,11], is a process used to join two metal parts heated locally: this can occur with or without addition of another metal, which is the filler metal [12,13], melted between the edges to be joined. The welding is heterogeneous when the molten filler material, which necessarily must have a lower melting point, shows a different composition from that of the pieces to be welded; this is the case of brazing in all its variants. Otherwise, the autogenous welding provides the fusion of the base metal (filler) which therefore must have a similar composition, or the fusion of only the edges to be welded pushed together by means of pressure (gas welding or electric arc) [14]. Brazing is carried out by placing the base metal so that, between the parts to be joined, there is a space both for allowing the filling of the joint and obtaining a connection for wetting and capillary action [15]. Depending on the lower or higher melting point of the filler metal, brazing stands in soft and hard. The soldering uses filler materials with melting temperature <450 °C; the filler materials are typical tin/lead alloys.

On the other hand, micro-welding is largely used in high-tech applications, especially for electronic devices [16,17]. The solder in electronics [18] has the main goal to ensure a good electrical connection between the parts to be welded [19,20,21]. It is also important and necessary to ensure a rigid mechanical, sturdy and long-lasting connection between the two metal surfaces. It is well to specify that the weld does not realize a simple 'bonding' but between the solder and the material it will form a thin transition layer consisting of an alloy composed of solder alloy and the material, the layer of which is called alloying. In fact, during the welding of Cu wire with Sn-Pb alloy, a layer of Cu-Sn-Pb is formed: it ensures good mechanical and electrical connections [22]. If this layer is not formed, the welding is poor (cold welding) and both the electrical and mechanical connections are poor. Micro-welding is based on different technologies and equipment: these have the scope to be able to perform welding on thin or delicate materials. The welding processes are designed to maintain tight operational tolerances and to not interfere with product function and quality. Most of the micro-welding operations are carried out under microscope with suitable miniature tools to place exactly the hardware elements and filler metal (if present), and to control performance [23]. Devices where components are welded need that the equipment used for micro welding is able to locate the hardware itself with the required precision. The required amount of energy for welding is deposited in the exact spot. Although micro-welding, also called precision welding, is not a standard term identifying a specific process, and recognized as such by the AWS Terms and Definitions Committee, it is a loose description of any process applicable to thin or tiny joints for the purpose of establishing a permanent connection. Also, there is not yet universally accepted consensus on size. A common definition considers micro welds as all those having fusion zone dimensions less than 100 μm. Another definition would accept welds less than 0.5 mm wide as micro welds. Practically, any welding process suitable to be miniaturized in order to accomplish required diminutive welding could be included in the micro welding term. The following processes have been applied with success to micro-welding: resistance, flash, arc (TIG, MIG and plasma) and laser [24]. Also, electron beam welding could be considered for miniature welding. Applications include tiny medical implants, lamps and lighting assemblies, jewelry, dental appliances, aerospace items, hermetic sealing, batteries, thermocouples, wires, TV tube parts, sensing devices, electronic instrumentation, molds and tool build up and many other types of miniature work [25].

This paper shows the evaluation of number particle size distribution and relative surface area of submicron aerosol during (micro-)welding operations and, in particular, our attention has been focused on the particles in the nucleation mode (5.6–16 nm): number concentration levels and behavior will be reported and debated according to the 'hard-welding' and 'soft-welding' investigated in this study. Further, a model describing the fate of such particles in the respiratory system in terms of deposition dose will be described: the evaluation of the deposited aerosol doses will be discussed in relationship with the particle diameter and the generations in the respiratory tract during welding operations. This information will be useful for understanding the risk of the workers exposed to welding (fumes) aerosol.

2. RESULTS AND DISCUSSION

Real-time samplings were performed during repeated operations of Sn-Pb micro-welding under suction. Figure 1a shows the trends of the temporal evolution of submicron particles (5.6–560 nm).

Firstly, it can be seen how the total number of particles produced is of the order of a million (over 10^6 # cm^{-3}) with a difference of 10^3 # cm^{-3} compared to the background level (around 10^3 # cm^{-3}). In detail, it is very interesting to note that particles in nucleation mode (5.6–16 nm) can be an important fraction: they are averagely 30% of the total fraction. Welding aerosols are initially formed from the nucleation of vapors emanating from the superheated metal droplets located within the arc, and from spatter particles ejected from the welding process [26]. The collision frequency of freshly nucleated particles with larger particles is high due to the Brownian diffusion enhanced by the high temperatures reached during welding, and to the high particle concentration. As a consequence, these particles are quickly formed and are removed to form larger particles, and aerosol size distributions quickly evolve with time. In case of repeated micro-weldings (basically, 8–10), when the dispersion of the particles is only limited by time factors, but permitted by a suction system, higher percentage ratios (up to 45%–50% of the total particles) are also reached. The micro-welding operation averagely produces 92000 # cm^{-3} (max 1.4×10^6 # cm^{-3}) particles in contrast to the average number of particles produced during industrial welding operations (about 3×10^5 # cm^{-3}, max 2×10^6 # cm^{-3}) [27].

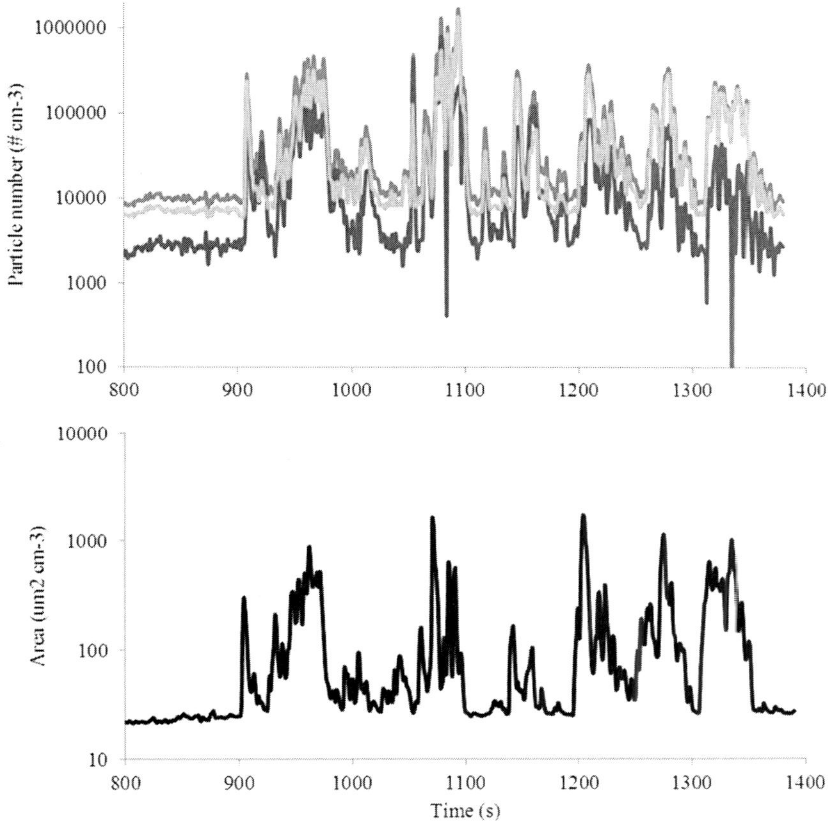

Figure 1. Performance of the three investigated modes (6–560 nm blue line, 5.6–16 nm red line, 19–560 nm green line) (**a**) and the particle surface area (**b**) during the Sn-Pb micro-welding operations.

Figure 1b shows the simultaneous trend of the surface area of the particles during the same operation reported in Figure 1a: also in this case, there is a periodicity due to the repetition of the operation carried out. The measurement of the surface area is an important indication from the point of view of human health: the parameter thus becomes fundamental in the study of the depositions (total deposited alveolar surface area 0.012 cm^2) along the respiratory tree with the relative doses. Stoeger *et al.* [28] showed that deposited surface area represents a valuable reference unit for the assessment of causative health effects for carbonaceous UFPs. Moreover, they demonstrated the existence of a threshold for the particle surface area at an instilled dose of approximately 20 cm^2, below which no acute proinflammatory responses could be detected in mice.

Figure 2a shows the behavior of the submicron particles during brazing operations. Such brazing operations are carried out in different environments than those where micro-welding operations previously analyzed are carried out: in this area, there are several sources of particles, external sources causing an elevated background level.

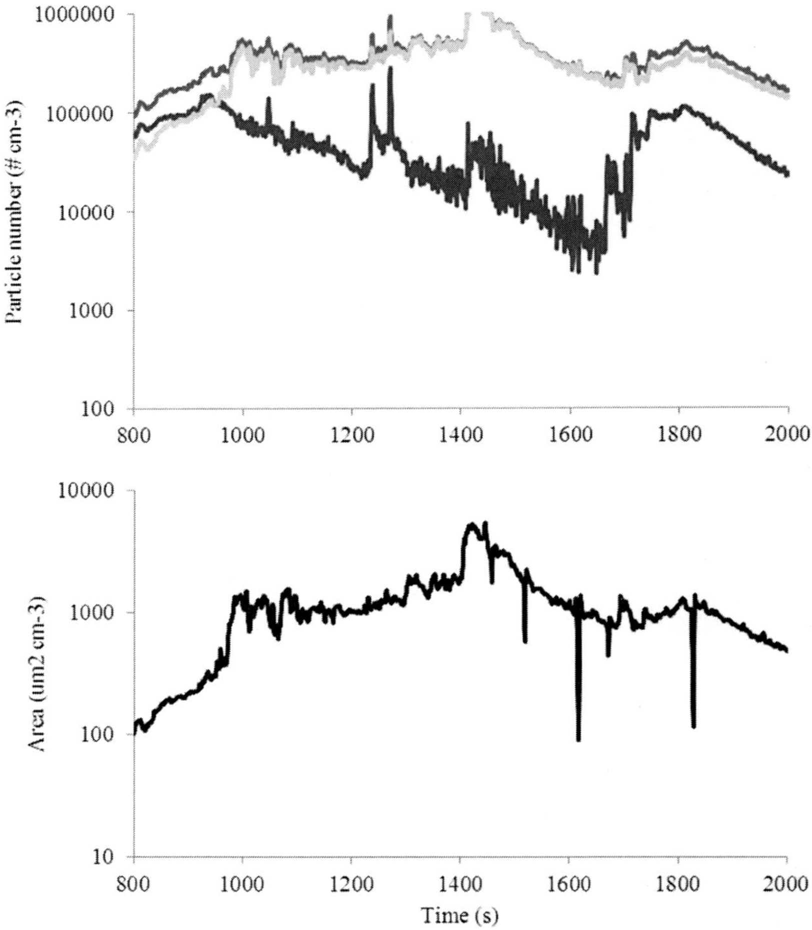

Figure 2. Performance of the three investigated modes (5.6–560 nm blue line, 5.6–16 nm red line, 19–560 nm green line) (**a**) and the particle surface area (**b**) during the Cu-Al soldering operations.

In such measures (Figure 2a), it can be noted how the background level is very high, around 1 ÷ 3 × 10^4 # cm^{-3}, whereas maximum particle peaks during similar operations are not so high. Moreover, particles in the nucleation mode exhibit different temporal trends: while the particles in the mode 5.6–16 nm decrease in number during the operations, and the particles in the mode 19–560 nm decrease, or at least remain constant, due to accumulation phenomena occurring in atmosphere. This can be explained by Figure 3 where particles in the nucleation mode are rapidly formed and are quickly removed by collision with accumulation mode particles.

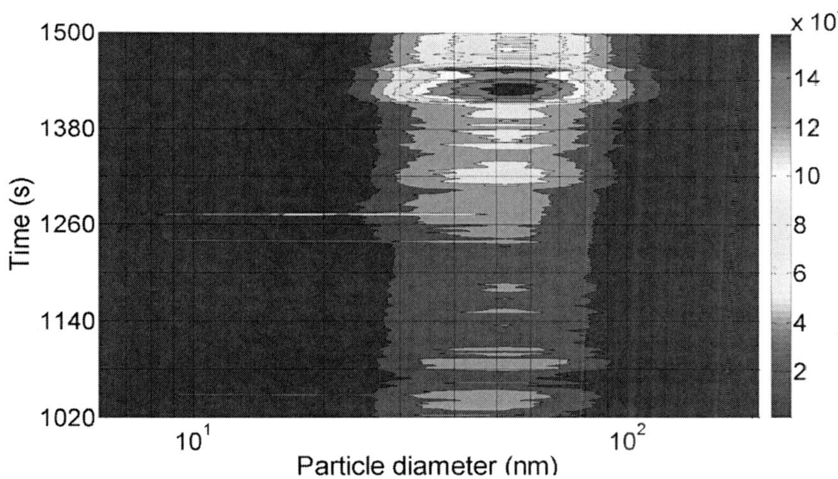

Figure 3. Temporal trend of aerosol size distribution (# cm^{-3}) during the Cu-Al soldering.

Figure 2b shows the trend of the particle surface area: a very different behavior than previously studied (total deposited alveolar surface area 0.245 cm^2) can be evidenced. In this case, the trend is continuously increasing, qualitatively very similar to the performance of the particles in the mode 19–560 nm with the exception of some peaks (negative) in correspondence with the operation change (further addition of borax on the sample to be brazed).

It is interesting to analyze the correlations between particles and surface area for the micro-welding and the brazing as well. Excellent correlation (r^2 0.95) is noted between the number of particles in the mode 5.6–560 nm and the surface area during the brazing process, whereas this correlation is not good for the micro-weldings (r^2 0.25): this is probably due to the different surface action employed during brazing and so the measurements are statistically more correlated. Inside of the measures for a same process, however, the correlations between the total particles (5.6–560 nm) and those in the nucleation mode (5.6–16 nm) or those in the accumulation mode (19–560 nm) are good for the micro-weldings (0.59 per particles in the nucleation mode and 0.96 for the particles in the accumulation mode) whereas show values contrasting for brazing (−0.29 for the particles in the nucleation mode and 0.99 for the particles in the accumulation mode).

Figure 4 shows the trend of total particles (5.6–560 nm) and particles in the mode 5.6–16 nm in the welding area during a typical MIG operation (electrode E 7018; heat input 0.43 kJ mm^{-1}).

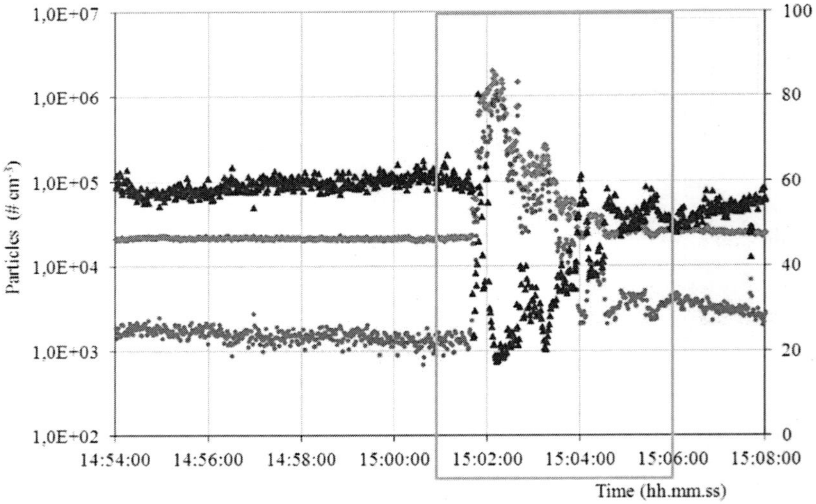

Figure 4. Submicronic particles (5.6–16 nm red circles; 5.6–560 green diamonds) and geometric mean diameter (blue triangles) during a Metal Inert Gas (MIG) welding.

The welding operation was carried out with suction whereas the sampling probe is positioned close to the respiratory system of the worker. This operation causes an instantaneous increase in the number of total particles and, in particular, ultrafine particles. Moreover, the considerations reported for Figure 1 as to nucleation mode particle can be easily seen in this figure where such particles steeply increase and become an important fraction of the total particle concentration (see box in Figure 4). In fact, the greatest increase is due to the particles with particle size 5.6–16 nm (about 3 orders of magnitude, up to 10^6 particles cm^{-3}), whereas the particles between 5.6 and 560 nm reach a maximum value over 2 × 10^6particles cm^{-3}. At the end of the welding operation, which lasted about 2 min, the particles in the range 5.6–16 nm employ a latency time of about 3 min for returning to their plateau value around 25,000 particles cm^{-3}. The proceeding of the welding particles in the size range 5.6–16 nm are greatly dispersed; this is due to the generation process in progress (welding) and the simultaneous removal, particularly efficient for these particles and due to their Brownian diffusion, and to the coagulation process after their collision with other particles. The particles in the mode 19–560 nm have, however, higher atmospheric residence times since both Brownian diffusion and the gravitational sedimentation are of poor efficiency for them.

After about 40 s from the start of welding, the particles in the nucleation mode increase from 7%–60%; simultaneously, geometric mean diameter (GMD) decreases from 60 nm to about 18 nm. The highlighted part of this figure provides an attempted estimated of the doses that are deposited in the respiratory tract.

It should also be considered that the toxicity of welding fumes depends both on their chemical composition and the ability to penetrate and accumulate in the lungs. Their penetration and deposition in regions of the respiratory system is determined mainly by the size of the particles. An important part of this study is the estimation of the dose of aerosol deposited in function of the particle diameter and the generations of the respiratory system for a worker.

Ultrafine particles are extremely small when compared with the cellular structures and have a high surface area per unit mass [29,30,31]. They form aggregates and are easily deposited as aggregates in the lungs, where it can then disaggregate. About this, it should be noted that the measurements performed in this study show high concentrations of particles with numerical size less than 1 µm, in particular those in the nucleation mode. There are studies documenting the adverse effects of ultrafine particles [32,33,34,35,36] while current research is seeking to understand those mechanisms [37,38,39]. In any case, the high surface area of these particles seems to lead to phenomena of oxidative stress, and changes in calcium levels in macrophages and epithelial cells (promoting inflammatory mechanisms). For such particles, the surface area, rather than the mass, is the measure of the dose that best predicts the lung response, with important consequences for the occupational standards currently based on the particle mass [40,41].

The choice of the type of welding with respect to another one derives essentially from the type of welding: the particle production and their evolution are shown in detail in Figure 5, where in highlight it is also shown the size distribution of the particles in the range 5.6–560 nm at the maximum peak (2×10^6 # cm^{-3}). In particular, from the analysis of the latter figure a bi-modal performance of the particles are evident with maximum peaks at 12 nm and 18 nm with values reaching 2.8×10^5 # cm^{-3} for the mode at 12 nm, whereas the particle number drastically decreases for particle sizes greater than 100 nm.

The possibility to monitor this rapidly evolution, reported in Figure 6, is fundamental in the study of dose deposition in the human respiratory tract. In this case the use of FMPS is critical because it allows for a high temporal resolution [42]: in fact, the size aerosol distribution is measured every second (1 Hz).

The assessment of the doses was performed by the MPPD model: it calculates the deposition and clearance of aerosol mono- and poly-disperse, from ultrafine particles (<100 nm) to coarse mode particles in the human and rat respiratory systems; it considers the efficiency of particle deposition by diffusion, sedimentation and impact.

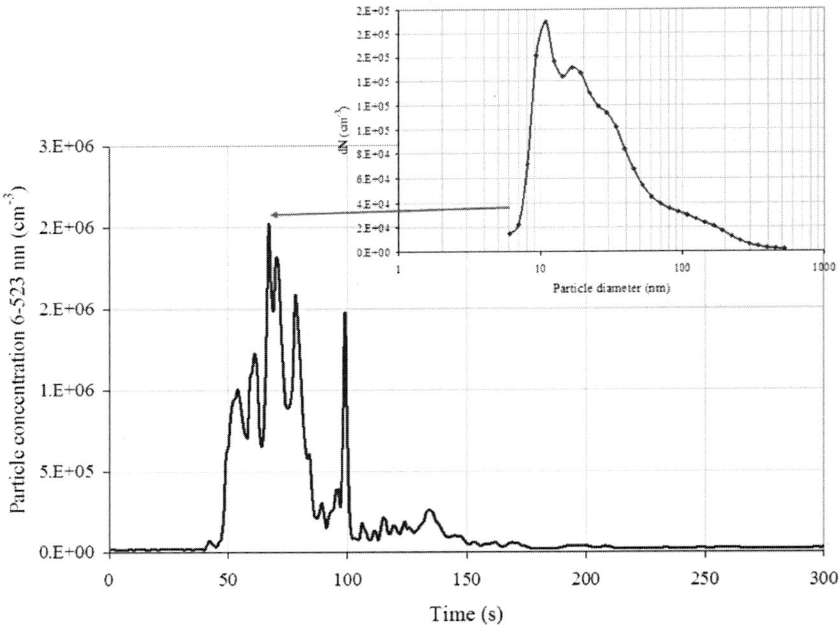

Figure 5. Evolution of the particle number concentration during an entire welding operation along with the highlight showing the particle number size distribution at the maximum particle generation peak.

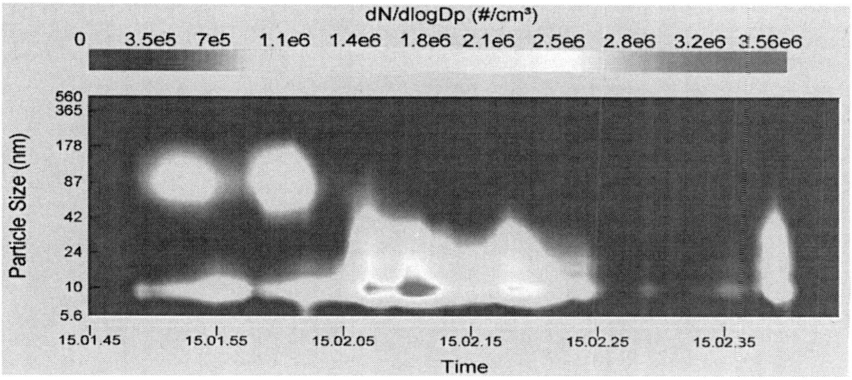

Figure 6. Dimensional distribution showing the rapid temporal evolution of particles during welding operations (Fast Mobility Particle Sizer software, release version 3.1.0.0, TSI, Shoreview, MN, USA).

In Table 1, the parameters of the model Multiple-Path Particle Dosimetry (MPPD) [43,44,45] used to evaluate the deposition of particles in the respiratory tract [46,47] during the welding operation described in Figure 4 are reported.

Furthermore, the doses are calculated as a function both of the particle size and the generations of the respiratory apparatus (28 considering the trachea as generation 0). Such model includes single-path and multi-path methods for the calculation of the airflow and the aerosol deposition. On the other hand, the modeling of the human respiratory tract is based on an asymmetric stochastic model of the lungs: five symmetrical lobes (three lobes on the right side, two lobes on the left side) but each lobe is structurally different.

Figure 7 shows the deposition in the respiratory tract during the welding: 9.8×10^9 particles are deposited in the respiratory tract after approximately 80 s while the maximum deposition is at the 21st generation (8.93×10^9 particle after 80 s).

Finally, Figure 8 shows the variations both of the dose deposited for each generation (red line) and the cumulative dose deposited (blue line) during a single welding process whereas the Figure 9 shows a contour plot where airway generation numbers are plotted *vs.* particle diameter. Analyzing both the graphs, it can be seen how the main contribution is due to the particles in the nucleation mode (maximum deposition for particles of 10.8 nm at the 20th generation, D_{max}, 1.36×10^7particles) while it is estimated that the density of deposition is the maximum at lobar bronchi (third generation) for particles of 10.8 nm (the maximum deposition density calculated overall size class-generation number data pairs, τ_{max}, is 8.40 × 10^4particles cm^{-2}).

Table 1. Multiple-Path Particle Dosimetry (MPPD) model parameters.

Stochastic Lung 60th Percentile	Data
Functional Residual Capacity (FRC)	3300 mL
Upper Respiratory Tract (URT) volume	50 mL
Particle density	3.7 g cm^{-3}
Breathing frequency	20 per minute
Tidal volume	1250 mL
Inspiratory fraction	0.5
Pause fraction	0
Breathing scenario	nasal

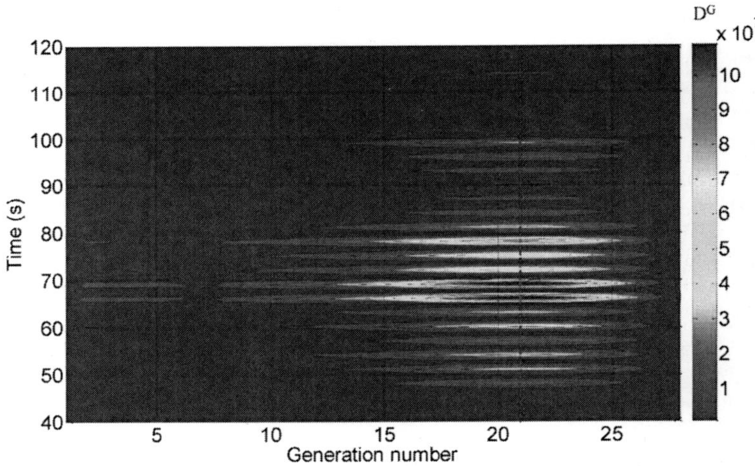

Figure 7. Contour plot showing the particle deposition for each generation number during a welding process: maximum deposition at the 21st generation 8.93 × 10^9particles (D^G: total particle number dose deposited in the Gth generation).

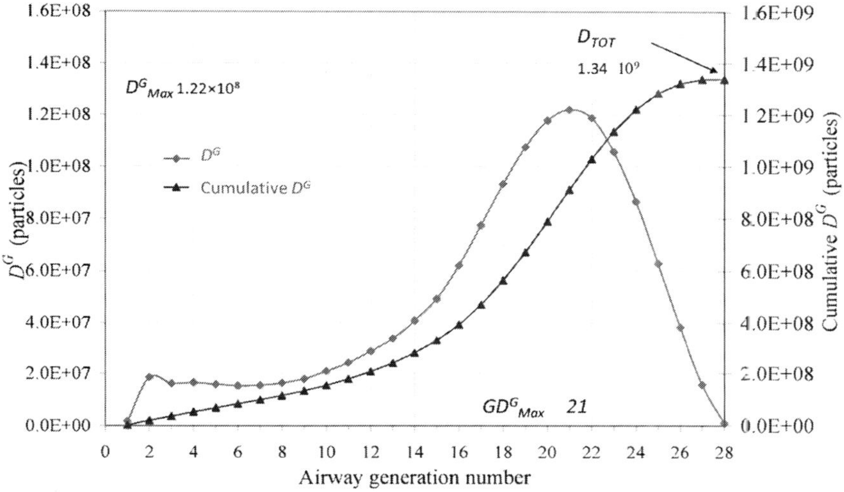

Figure 8. Dose deposited (**red**) for each generation and cumulative deposited dose (**blue**) during a single welding process (D^G the total particle number dose deposited in the Gth generation; $D^G{}_{max}$ the maximum dose per airway generation; D_{tot} the total particle number dose deposited in the respiratory tree; $GD^G{}_{max}$ the airway generation number at $D^G{}_{max}$).

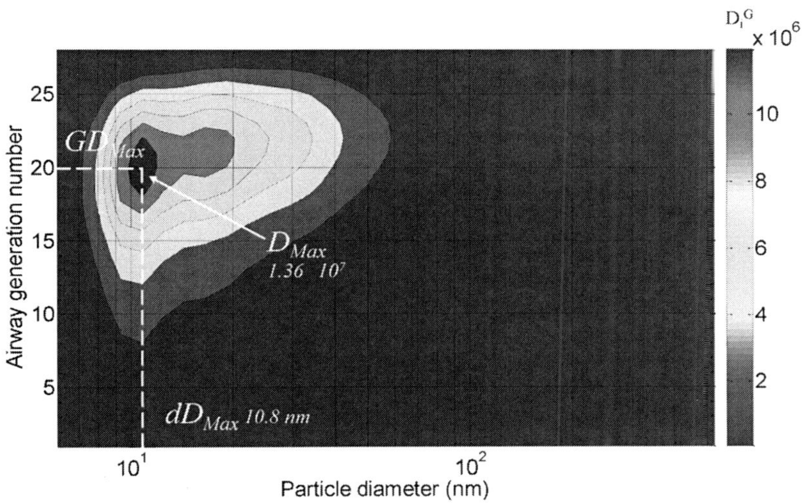

Figure 9. Contour plot reporting the airway generation number in relationship with the particle diameter ($D^G{}_i$ the total particle number dose deposited in the i-generation;$GD^G{}_{max}$ the airway generation number at $D^G{}_{max}$).

3. EXPERIMENTAL SECTION

3.1. Materials and Methods

This paper investigates the temporal evolution of submicron particles in the range 6–523 nm particle size (calculated as electric mobility diameter) and in particular the particle behavior in a size range (5.6–16 nm) that is representative of the nucleation mode. All the experimental measurements are started before the holding of the welding operations and are terminated after several minutes (between 10 and 20 min, in general) so as to restore the background operating conditions of the acquisition system. The measurements were carried out both in the presence and absence of suction and ventilation devices. This measure has been used the Fast Mobility Particle Sizer TM Spectrometer (Model 3091 FMPS, TSI, Shoreview, MN, USA), a well-known instrument with high resolving size and time resolution, which allows collection, counting and provide the size distribution of particles with size from 5.6–560 nm [48,49]. Unlike the Scanning Mobility Particle Sizer (SMPS) [50,51] sampling takes place every second [52,53].

This instrument charges the particles due to collision with an ionized gas; the particles thus charged are separated in function of their electrical mobility through flowing into a further electric field generated by electrometers. The instrument has 32 channels used for the separation of the particles. This tool, one of the most technologically advanced in the field of aerosol, allows to obtain the particle size distribution in the entire dimensional spectrum explored in a very short time (1 s). This occurrence allows to observe the ultrafine particles produced by combustion phenomena with respect to the particles formed by agglomerations of smaller particles (coagulation and/or gas-particle condensation), following simultaneously the kinetics of formation.

Over the FMPS instrument, a Nanoparticle Surface Area Monitor (Model 3550, TSI, Shoreview, MN, USA) was used for measuring the human lung deposited surface area of particles (expressed as $\mu m^2\ cm^{-3}$) corresponding to tracheobronchial (TB) and alveolar (A) regions of the lung depending on the ion trap voltage [54]. The NSAM provides a simple and fast solution for detecting the surface area equivalent dose in the lung of particles in the size range of 10 to 1000 nanometers. The instrument is based on diffusion charging of sampled particles, followed by detection of the charged aerosol using an electrometer. The aerosol sample is drawn into the NSAM passing through a cyclone with 1 μm cut point. The aerosol moves on to the electrometer for charge measurement. In the electrometer, current is passed from the particles to a conductive filter and measured by a very sensitive amplifier. The charge measured by the electrometer is directly proportional to the surface area of the particles passing through the electrometer The ion trap voltages are optimized to correspond to the ICRP model-based tracheobronchial and alveolar lung deposition curves [55], and the NSAM indicates the lung deposited surface area, not the total active surface area of the aerosol sampled.

3.2. Welding and Brazing Operations

The measurements, repeated in several stages have been performed in different work processes: in the assembly, micro weldings of electronic circuits and/or connections have been carried while in the brazing area solderings were performed with brazing filler metal without melting the metal welding (Figure 10). The sampling probe was located on the worker's lapel.

In Pb and Sn micro-welding, the process took place under vacuum system (suction Weller WFE 2ES, Sparks, MS, USA; suction flow conveyed in a closed container with, in series, a pre-filter to reduce particulate and an activated carbon filter). The entire working operation of the micro-welding has occurred at the tip temperature of 320 °C (with a peak of 380 °C); the thread of Sn/Pb (ratio 60/40) is of Sn section 0.5 mm. During the soldering/desoldering operations, a solution of alcohol and chloroform was used as a fluxing.

After, the exposure was evaluated in the brazing process. The material to be welded is Cu or Al, welded with filler material that is castolin (based-silver alloy) and alloy Al-Si, respectively. Aspiration is performed from the bottom (carbon filters, SAF-FRO Saldidactic Professional Welding Fumes Electrostatic Filtering System, Air Liquide Welding, Cergy-Pontoise, France); the temperature reached is of about 800 °C. During the brazing operation borax is used. The welding torch is based on an oxy-acetylene flame. In this type of process (strongly dependent on the demand of the market), the operator is involved an average of 10 days per month with an average of 4 h per day.

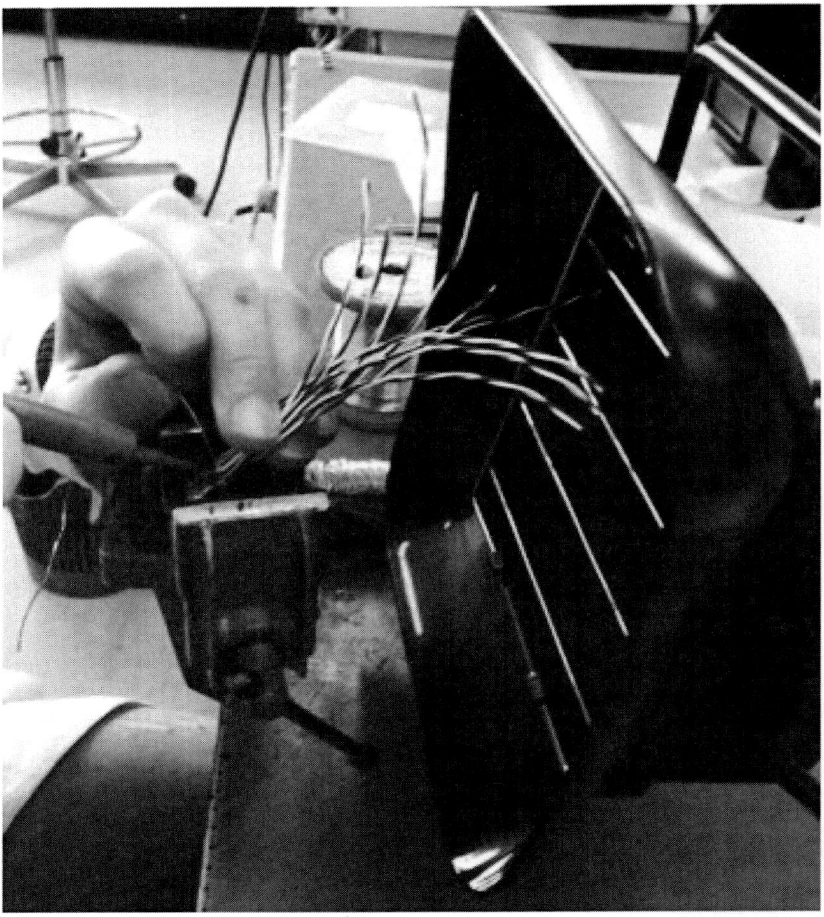

Figure 10. The micro-welding operation under suction (sampling probe was located on the worker's lapel).

4. CONCLUSIONS

The results of this study confirm that in the engineering industry and, particularly, during welding operations, there is significant production of submicronic (and ultrafine too) particles. Due to their size (<100 nm), they manage to reach the deeper portions of the respiratory system, spreading into the alveoli and reaching the bloodstream.

The analysis of the temporal distributions of particle size also highlights the following points:

- very fast evolution of the particle size in the range of 5.6–560 nm during welding operations, in particular those in the nucleation mode;
- importance of measures at high temporal resolution (1 s) performed by FMPS for evaluating the particle size distributions during generation;
- the total deposited alveolar surface area values are 20-fold higher for Cu-Al soldering than the Sn-Pb micro-welding. To the authors' knowledge, studies correlating surface area to health effects have not been performed on metal fumes; nonetheless, the data obtained in this study represent a useful reference to interpret the health effects observed for welders;
- due to their high particle deposition efficiency in the human respiratory system, such particles pose a (potential) risk to the human health associated with their high number, concentration and/or surface area. They should be the subject of further studies to continuously evaluate and monitor their effects.

ACKNOWLEDGMENTS

This research was performed under the grants INAIL/P20L01 and INAIL/P20L09.

AUTHOR CONTRIBUTIONS

Pasquale Avino, Pietro Pandolfi and Nicola Paolucci are responsible for the research design. Pasquale Avino, Pietro Pandolfi, Cosimo Tornese and Diego Settimi conducted the field work. The paper was written by Pasquale Avino with significant contributions by Maurizio Manigrasso and Pietro Pandolfi. All the authors approved the paper.

CONFLICTS OF INTEREST

The authors declare no conflict of interest.

REFERENCES

1. Avino, P.; Manigrasso, M.; Carrai, P.; Rocchi, C.; Guerriero, E.; Russo, M.V. Ultrafine particles and chemical risk in automobile repair shops. *Fresenius Environ. Bull.* **2014**, *23*, 2956–2966.
2. Howe, A.M. Assessment of Exposure to Chemical Agents in Welding Fume: Final Report. HSL/2000/15. 2000. Available online: http://www.hse.gov.uk/research/hsl_pdf/2000/hsl00-15.pdf (accessed on April 2015).
3. Riediger, G.; Möhlmann, C. Ultrafeine aerosole an arbeitsplätzen. *Gefahrstoffe* **2001**, *61*, 429–434.
4. Antonini, J.M. Health effects of welding. *Crit. Rev. Toxicol.* **2003**, *33*, 61–103.
5. Antonini, J.M.; Taylor, M.D.; Zimmer, A.T.; Roberts, J.R. Pulmonary responses to welding fumes: Role of metal constituents. *J. Toxicol. Environ. Health Part A* **2004**, *67*, 233–249.
6. Pires, I.; Quintino, L.; Miranda, R.M.; Gomes, J.F.P. Fume emissions during gas metal arc welding. *Toxicol. Environ. Chem.* **2006**, *88*, 385–394.
7. Gomes, J.F.P.; Albuquerque, P.C.S.; Miranda, R.M.M.; Vieira, M.T.F. Determination of airborne nanoparticles from welding operations. *J. Toxicol. Environ. Health Part A* **2012**, *75*, 747–755.
8. Maynard, A.D.; Zimmer, A.T. Evaluation of grinding aerosols in terms of alveolar dose: The significance of using mass, surface area and number metrics. *Ann. Occup. Hyg.* **2002**, *46*, 315–319.
9. Jenkins, N.T.; Eagar, T.W. Chemical analysis of welding fume particles. *Weld. J.* **2005**, *84*, 87s–93s.
10. Timings, R. *Fabrication and Welding Engineering*, 1st ed.; Elsevier: Oxford, UK, 2008; pp. 467–519.
11. Messler, R.W., Jr. *Principles of Welding—Processes, Physics, Chemistry, and Metallurgy*, 1st ed.; Wiley-VCH Verlag: Weinheim, Germany, 2004; pp. 3–40.

12. Kim, K.H. *Welding Handbook Filler Materials for Manual and Automatic Welding*, 5th ed.; Esab AB: Göteborg, Sweden, 1991; pp. 123–146.

13. O'Brien, R. *Jefferson's Welding Encyclopedia*, 18th ed.; American Welding Society: Miami, FL, USA, 1997; pp. 683–684.

14. Khan, I. *Welding Science and Technology*, 1st ed.; New Age International (P) Ltd.: New Delhi, India, 2007; pp. 37–68.

15. Schwartz, M.M.; Aircraft, S. Introduction to brazing and soldering. In *ASM Handbook—Welding, Brazing, and Soldering vol. 6*, 9th ed.; Olson, D.L., Siewert, T.A., Liu, S., Edwards, G.R., Eds.; American Society for Metals International: Materials Park, OH, USA, 1983; pp. 270–481.

16. Zhou, Y. *Microjoining and Nanojoining*, 1st ed.; Woodhead Publishing Ltd.; Woodhead Publishing: Cambridge, UK, 2008; pp. 1–810.

17. Fukumoto, S.; Tsubakino, H.; Zhou, Y. Resistance microwelding of fine nickel wires. In ASM Conference Proceedings: Joining of Advanced and Specialty Materials VII, Proceedings of the Conference on Materials Solutions 2004 on Joining of Advanced and Specialty Materials, Columbus, OH, USA, 18–20 October 2004; ASM International: Materials Park, OH, USA, 2005; pp. 168–173.

18. Pawlak, R.; Kostrubiec, F.; Tomczyk, M.; Walczak, M. Laser-induced change of electrical resistivity of metals and its applications. In Proceedings of SPIE—The International Society for Optical Engineering, Proceedings of the Conference on Lasers in Material Processing and Manufacturing II, Beijing, China, 10 November 2004; The International Society for Optical Engineering: Bellingham, WA, USA, 2005; Article number 55. pp. 349–360.

19. Fendrock, J.J.; Hong, L.M. Parallel-gap welding to very-thin metallization for high temperature microelectronic interconnects. *IEEE Trans. Compon. Hybrids Manuf. Technol.* **1990**, *13*, 376–382.

20. Steinmeier, D. Downsizing in the world of resistance welding. *Weld. J.* **1998**, *77*, 39–47.

21. Ely, K.J.; Zhou, Y. Microresistance spot welding of Kovar, steel, and nickel. *Sci. Technol. Weld. Join.* **2001**, *6*, 63–72.

22. Mo, B.; Guo, Z.; Li, Y.; Huang, Z.; Wang, G. Mechanism of resistance microwelding of insulated copper wire to phosphor bronze sheet. *Mater. Trans.* **2011**, *52*, 1252–1258.

23. Fukumoto, S.; Zhou, Y. Mechanism of resistance microwelding of crossed fine nickel wires. *Metall. Mater. Trans. A Phys. Metall. Mater. Sci.* **2004**, *35*, 3165–3176.

24. Ventrella, V.A.; Berretta, J.R.; de Rossi, W. Application of pulsed Nd: YAG laser in thin foil microwelding. *Int. J. Mater. Product Technol.* **2014**, *48*, 194–204.

25. Ely, K.J.; Hall, P.; Zhou, Y. Microwelding methods in medical components and devices. In *Joining and Assembly of Medical Materials and Devices*, 1st ed.; Zhou, Y.N., Breyen, M., Eds.; Woodhead Publishing Series in Biomaterials: Cambridge, UK, 2013; pp. 47–78.

26. Zimmer, A.T.; Biswas, P. Characterization of the aerosols resulting from arc welding processes. *Aerosol Sci.* **2001**, *32*, 993–1008.

27. Avino, P.; Manigrasso, M.; Fanizza, C.; Carrai, P.; Solfanelli, L. Particelle submicrometriche in fumi derivanti da operazioni di saldatura e di fusione di leghe metalliche. *La Med. del Lav.* **2013**, *104*, 335–350.

28. Stoeger, T.; Reinhard, C.; Takenaka, S.; Schroeppel, A.; Karg, E.; Ritter, B.; Heyder, J.; Schulz, H. Instillation of six different ultrafine carbon particles indicates a surface area threshold dose for acute lung inflammation in mice.*Environ. Health Perspect.* **2006**, *114*, 328–333.

29. Oberdörster, G. Pulmonary effects of inhaled ultrafine particles. *Int. Arch. Occup. Environ. Health* **2001**, *74*, 1–8.

30. Kreyling, W.; Semmler, M.; Mayer, P.; Takenaka, S.; Schulz, H.; Oberdörster, G.; Ziesenis, A.; Erbe, F. Translocation of ultrafine insoluble iridium particles from lung epithelium to extrapulmonary organs is size dependent but very low. *J. Toxicol. Environ. Health Part A* **2002**, *65*, 511–535.

31. Avino, P.; Casciardi, S.; Fanizza, C.; Manigrasso, M. Deep investigation of ultrafine particles in urban air. *Aerosol Air Qual. Res.* **2011**, *11*, 654–663.

32. Pourtaghi, G.H.; Kakooei, H.; Salem, M.; Pourtaghi, F.; Lahmi, M.A. Pulmonary effects of occupational exposure to welding fumes. *Aust. J. Basic Appl. Sci.* **2009**, *3*, 3291–3296.

33. Fanizza, C.; Casciardi, S.; Avino, P.; Manigrasso, M. Measurements and characterization by Transmission Electron Microscopy of ultrafine particles in the urban air of Rome. *Fresenius Environ. Bull.* **2010**, *19*, 2026–2032.

34. Lehnert, M.; Pesch, B.; Lotz, A.; Pelzer, J.; Kendzia, B.; Gawrych, K.; Heinze, E.; van Gelder, R.; Punkenburg, E.; Weiss, T.; *et al.* Exposure to inhalable, respirable, and ultrafine particles in welding fume. *Ann. Occup. Hyg.* **2012**, *56*, 557–567.

35. Occupational Safety and Health (OSHA). Occupational Safety and Health Guideline for Welding Fumes. Available online: http://www.osha.gov/SLTC/healthguidelines/weldingfumes/recognition.html (accessed on March 2015).

36. Oberdörster, G.; Finkelstein, J.N.; Johnston, C.; Gelein, R.; Cox, C.; Baggs, R.; Elder, A.C. Acute pulmonary effects of ultrafine particles in rats and mice. *Res. Rep. Health Eff. Inst.* **2000**, *96*, 5–74.

37. Oberdörster, G. Toxicology of ultrafine particles: *In vivo* studies. *Philos. Trans. R. Soc. B* **2000**, *358*, 2719–2740.

38. Donaldson, K.; Stone, V.; Seaton, A.; MacNee, W. Ambient particle inhalation and the cardiovascular system: Potential mechanisms. *Environ. Health Perspect.* **2001**, *109*, 523–527.

39. Oberdörster, G.; Oberdörster, E.; Oberdörster, J. Nanotoxicology: An emerging discipline evolving from studies of ultrafine particles. *Environ. Health Perspect.* **2005**, *113*, 823–840.

40. Buonanno, G.; Morawska, L.; Stabile, L. Exposure to welding particle in automotive plants. *J. Aerosol Sci.* **2011**, *42*, 295–304.

41. Gomes, J.; Albuquerque, P.; Miranda, R.; Santos, T.; Vieira, M. Comparison of deposited surface area of airborne ultrafine particles generated from two welding processes. *Inhal. Toxicol.* **2012**, *24*, 774–781.

42. Manigrasso, M.; Avino, P. Fast evolution of urban ultrafine particles: Implications for deposition doses in the human respiratory system. *Atmos. Environ.* **2012**, *51*, 116–123.

43. Manigrasso, M.; Stabile, L.; Avino, P.; Buonanno, G. Influence of measurement frequency on the evaluation of short-term dose of sub-micrometric particles during indoor and outdoor generation events. *Atmos. Environ.* **2013**, *67*, 130–142.

44. Manigrasso, M.; Buonanno, G.; Fuoco, F.C.; Stabile, L.; Avino, P. Aerosol deposition doses in the human respiratory tree of electronic cigarette smokers. *Environ. Pollut.* **2015**, *196*, 257–267.

45. Manigrasso, M.; Buonanno, G.; Stabile, L.; Morawska, L.; Avino, P. Particle doses in the pulmonary lobes of electronic and conventional cigarette users. *Environ. Pollut.* **2015**, *202*, 24–31.

46. Stabile, L.; Buonanno, G.; Avino, P.; Fuoco, F.C. Dimensional and chemical characterization of airborne particles in schools: respiratory effects in children. *Aerosol Air Qual. Res.* **2013**, *13*, 887–900.

47. Avino, P.; Lopez, F.; Manigrasso, M. Regional deposition of submicrometer aerosol in the human respiratory system determined at 1-s time resolution of particle size distribution measurements. *Aerosol Air Qual. Res.* **2013**, *13*, 1702–1711.

48. Manigrasso, M.; Avino, P.; Fanizza, C. Ultrafine particles in the urban area of Rome. *Fresenius Environ. Bull.* **2009**, *18*, 1341–1347.

49. Buonanno, G.; Bernabei, M.; Avino, P.; Stabile, L. Occupational exposure to airborne particles and other pollutants in an aviation base. *Environ. Pollut.* **2012**, *170*, 78–87.

50. MacNee, W.; Li, X.Y.; Gilmour, S.; Donaldson, K. Systemic effects of PM10. *Inhal. Toxicol.* **2000**, *12*, 233–244.

51. Donaldson, K.; Stone, V.; Clouter, A.; Renwick, L.; Mac-Nee, W. Ultrafine particles. *Occup. Environ. Med.* **2001**, *58*, 211–216.

52. Buonanno, G.; Stabile, L.; Avino, P.; Vanoli, R. Dimensional and chemical characterization of particles at a downwind receptor site of a waste-to-energy plant. *Waste Manag.* **2010**, *30*, 1325–1333.

53. Buonanno, G.; Stabile, L.; Avino, P.; Belluso, E. Chemical, dimensional and morphological ultrafine particle characterization from a waste-to-energy plant. *Waste Manag.* **2011**, *31*, 2253–2262.

54. Marini, S.; Buonanno, G.; Stabile, L.; Avino, P. A benchmark for numerical scheme validation of airborne particle exposure in street canyons. *Environ. Sci. Pollut. Res.* **2015**, *22*, 2051–2063.

55. International Commission on Radiological Protection (ICRP). *Human Respiratory Tract Model for Radiological Protection. International Commission on Radiological Protection (ICRP), Publication 66*; Elsevier Science: Oxford, UK, 1994.

3

CHAPTER

Defect Detection in Arc-Welding Processes by Means of the Line-to-Continuum Method and Feature Selection

P. Beatriz Garcia-Allende *, Jesus Mirapeix, Olga M. Conde, Adolfo Cobo and Jose M. Lopez-Higuera

Photonics Engineering Group, University of Cantabria, Avda. de los Castros S/N, 39005 Santander, Spain

ABSTRACT

Plasma optical spectroscopy is widely employed in on-line welding diagnostics. The determination of the plasma electron temperature, which is typically selected as the output monitoring parameter, implies the identification of the atomic emission lines. As a consequence, additional processing stages are required with a direct impact on the real time performance of the technique. The line-to-continuum method is a feasible alternative spectroscopic approach and it is particularly interesting in terms of its computational efficiency. However, the monitoring signal highly depends on the chosen emission line. In this paper, a feature selection methodology is proposed to solve the uncertainty regarding the selection of the optimum spectral band, which allows the employment of the line-to-continuum method for on-line welding diagnostics. Field test results have been conducted to demonstrate the feasibility of the solution.

Keywords: arc-welding; plasma spectroscopy; feature selection; on-line monitoring

1. INTRODUCTION

The fields where welding processes are key factors in the production stages cover a wide range of different applications: manufacturing of pipes for various sectors, engines for aeronautics, automobiles or heavy components for nuclear power stations are some relevant examples in this regard. The lack of a complete comprehension of the physical phenomena occurring during the welding process, and the demanding quality standards to be found in this framework, have forced scientists to carry out an intense research effort in both welding physics and procedures devoted to cope with quality issues. Some of these studies have been focused on the development of theoretical models for both arc and laser welding [1-3], including numerical analysis approaches [4]. These efforts help to understand the process and, therefore, to determine the precise input parameter ranges that will provide seams free of flaws.

However, in practice welding coupons employed for parameter adjustment, and both destructive and non-destructive trials [5] have to be used to ensure that the performed seams satisfy the established quality standards. This obviously implies a significant cost in terms of productivity, as a lot of time is spent before and after the welding process itself, and, therefore, some of the seams have to be reworked and evaluated again.

This scenario has led to an intense research effort aimed at developing efficient and reliable on-line welding quality monitoring systems. They should be able to detect in real-time the occurrence of possible defects and, as an added value, to control the welding setup to try to avoid these defects or drifts from the standard operation conditions. Several techniques have been proposed, from electrical and capacitive sensors [6,7], to monitoring based on the analysis of the acoustic signal generated during the process [8,9] or solutions based on machine vision [10,11]. Among these alternatives, the optical analysis of the welding plasma radiation has proved to be a feasible and promising option. Initial proposals were based on the use of photodiodes and the analysis of emissions in the ultraviolet, visible and infrared regions [12], determining for example the full-penetration condition in laser welding [13].

A more sophisticated approach has been proposed by considering plasma optical spectroscopy, where emission lines appearing in the plasma spectra are analyzed to provide a plasma electron temperature T_e profile that shows a direct correlation to weld defects [14,15]. In the last years, several publications have dealt with refinements of this technique, allowing automatic defect detection [16] and reducing the overall computational cost of the system [17]. More recently, new strategies have been proposed to extract more information from the plasma spectra, like the correlation analysis proposed by Sibillano *et al.* [18], or proposals based on the use of optimization algorithms to generate synthetic spectra [19]. Within the same framework, new spectroscopic parameters are also being studied in an attempt to improve the monitoring system efficiency [20].

One of the key issues when using plasma spectroscopy lies in the correct selection of the emission lines chosen to calculate the output monitoring parameter. On the one hand, and depending on the selected instrumentation, there can be ambiguities on the emission line identification, what can end in unexpected results. On the other hand, and especially when defect classification is required, i.e., to be able to distinguish among different types of defects, it would be highly interesting to know which emission lines allow a better discrimination for classification purposes.

We have conducted some previous studies by using PCA (Principal Component Analysis) and SFFS (Sequential Forward Floating Selection) to feed an Artificial Neural Network [21,22]. The use of SFFS allows to gain knowledge about the best spectral bands selected. This will be used in this paper to propose a scheme based on both the SFFS algorithm and the line-to-continuum method [23] to generate the required output monitoring profiles. The line-to-continuum method implies the use of only a single emission line that, in addition, does not need to be identified, i.e., associated with its chemical species.

2. PLASMA OPTICAL SPECTROSCOPY FOR WELDING DIAGNOSTICS

The plasma electron temperature has been widely used as the output monitoring parameter for welding diagnostics, given the known correlation between its profiles and the appearance of defects in the seams. There are basically two approaches that are employed in the literature: a precise estimation of T_e can be obtained with the Boltzmann-plot method [23]:

$$\ln\left(\frac{I_{mn}\lambda_{mn}}{A_{mn}g_m}\right) = \ln\left(\frac{hcN}{Z}\right) - \frac{E_m}{kT_e} \tag{1}$$

where several emission lines from the same species are involved in the calculations. In the previous equation I_{mn} is the relative intensity of the chosen emission line, m and n the upper and lower states, respectively, λ_{mn} the central wavelength associated with the line, A_{mn} the transition probability, g_m the statistical weight, h the Planck's constant, c the light velocity, N the population density of the state m, Z the partition function, E_m the upper level energy and k the Boltzmann constant. T_e can be obtained if the left-hand side of Equation (1) is represented versus E_m, given that the slope of the resulting line is inversely proportional to the temperature.

On the other hand, and due to considerations regarding the computational performance of the monitoring system, which determines its spatial resolution, a simplification of the Boltzmann-plot method, where only two emission lines are involved, is typically used:

$$T_e = \frac{E_m(2) - E_m(1)}{k \ln\left[\dfrac{E_m(1) I(1) A(2) g_m(2) \lambda(1)}{E_m(2) I(2) A(1) g_m(1) \lambda(2)}\right]} \tag{2}$$

This equation was proposed by Marotta [24] for arc-welding processes. The commented techniques can be equally applied to both arc and laser processes, although for the latter the energies of the upper level disappear from the logarithm in the denominator.

Although different approaches can be taken into account, a possible processing scheme designed to provide the required T_e estimation from the acquired welding plasma spectra is presented in Figure 1.

The identification of the emission lines is compulsory to obtain T_e. As shown in Figure 1, this requires three additional processing stages (peak detection and line modeling and identification) and it has, as a consequence, a direct implication in the real-time performance of the overall approach. An alternative solution is to perform a previous spectral band selection stage with a data set consisting of spectra from the same welding process under different conditions. Afterwards those lines can be used without involving the identification in the processing scheme.

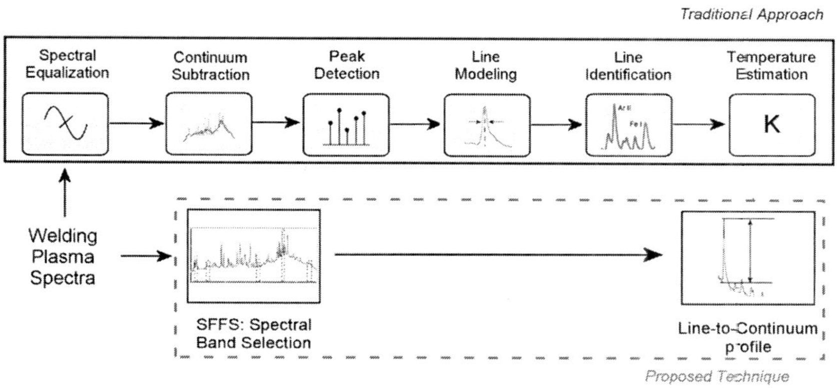

Figure 1. Comparison of the processing schemes associated with the traditional approach and the proposed solution based on SFFS and the line-to-continuum method.

This could be applied for scenarios where the same materials and welding conditions are used, but it limits the flexibility of the analysis strategy. On the other hand, as previously commented, there is a lack of knowledge on the selection of the optimal emission lines for welding diagnostics. Some studies have been carried out comparing the response of different elements and species, but we believe that by specifically searching for the most discriminant spectral bands the overall performance of the monitoring system should be improved.

Within this framework, the use of the line-to-continuum method to generate the output monitoring profiles seems to be a good solution, given that it does not require the identification of the chosen emission line. However, it could be performed to avoid problems related to the effect of unresolved lines. This method was originally intended to estimate T_e by means of the following expression [25]:

$$\frac{\varepsilon_l}{I_c}(\lambda) = 2.0052 \times 10^{-5} \frac{A_{mn} g_m}{Z_i} \frac{1}{T_e \xi} \exp\left(\frac{E_i - E_m}{kT_e}\right) \frac{\lambda}{\Delta\lambda} \quad (3)$$

where ε_l is the line intensity integrated over the line profile, I_c the intensity of the adjacent background radiation (non-integrated), Z_i is the ion partition function, ζ the free-bound continuum correction, E_i the ionization potential and $\Delta\lambda$ the wavelength bandwidth. It is worth mentioning that an iterative method has to be employed to determine T_e via Equation (3). However, in the proposed method we only use the ε_l / I_c ratio as the monitoring parameter. In a previous paper, this approach was initially explored in comparison to an alternative method based on the estimation of the wavelength associated with the maximum intensity of the continuum radiation [20]. In this case we concluded that the line-to-continuum method could not be reliably used given the uncertainty regarding the selection of the spectral band for the subsequent analysis. In this paper the use of the SFFS algorithm will help to deal with this issue.

3. SEQUENTIAL FLOATING FORWARD SELECTION OF SPECTRAL BANDS

The Sequential Floating Forward Selection (SFFS) algorithm [26] is widely applied to reduce the dimensionality, i.e., the number of features or wavelengths, of spectral data prior to their interpretation [27,28]. The spectral band selection criterion is based on the capability of the distinct features to separate the different classes to be discriminated afterwards. The greater the separability between the classes that a wavelength provides, the better the wavelength is for classification purposes. Therefore, a similar approach could be followed to solve the uncertainty encountered in the selection of the optimum band within the plasma spectra for on-line welding quality monitoring by means of the line-to-continuum method described above [20]. The aim of SFFS is to select M spectral bands that best discriminate among correct welds and flaws, out of the total number of initial bands N, so $M \ll N$. In this way the line-to-continuum profiles obtained later on for these M bands will be optimized for defect detection. It is worth highlighting here that SFFS has already been used in some previous works by our group [22]. However, the focus was, on the contrary, to provide a set of spectral bands to feed an Artificial Neural Network (ANN) for flaw discrimination as an alternative to the widely known feature extraction techniques [21].

As mentioned before, the selection criterion for band selection is an objective function based on a measurement of the separability of the classes. Specifically, it is estimated in terms of the Bhattacharyya distance as in [22]:

$$J_B = \frac{1}{4}(\mu_2 - \mu_1)^T \left[\Sigma_1 + \Sigma_2\right](\mu_2 - \mu_1) + \frac{1}{2}\log\left(\frac{\left|\Sigma_1 + \Sigma_2\right|}{2\left(\left|\Sigma_1\right| \cdot \left|\Sigma_2\right|\right)^{\frac{1}{2}}}\right) \tag{4}$$

where μ_i is the mean of the i class; Σ_i its covariance matrix and $|\Sigma_i|$ stands for the determinant of matrix Σ_i There are only two classes to be distinguished here: correct seams and defects. Therefore, the overall separability among the classes to be discriminated is straightly given by (4).

In Figure 2 several spectral curves of correct welds (a) and defects (b) are presented. As shown, some ranges of the spectra exhibit similar characteristics and, as a consequence, they are useless for defect detection purposes. The elimination of these ranges is compulsory prior to the application of SFFS. This procedure is named data decorrelation since only one spectral band, the one that maximizes the Bhatacharya distance, is selected within each redundant block. Each redundant block is obtained as a wavelength range where spectral bands have a high correlation coefficient, nearly 1.

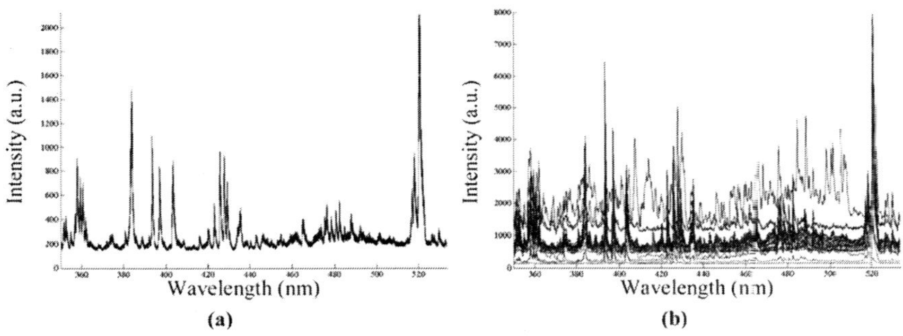

Figure 2. Spectral curves of: **(a)** several correct seams. **(b)** distinct flaws.

SFFS is named *"forward"* because it begins with an empty selected feature subset. It initially selects from the uncorrelated subset of spectral bands the most discriminant one according to the Bhattacharyya distance. Then it carries on sequentially adding the second most discriminant wavelength and so on. Therefore, in each iteration a new spectral band is added. However, if the iteration number is greater than 1, some of the previously added features can be removed if the separability of the classes enhances. Then, after every forward step (addition of a new spectral band) a number of backward steps (removal of several spectral bands) are applied as long as the resulting subsets are better than the previously evaluated ones according to the Bhatacharrya distance, and for that reason the selection procedure is also named "*floating*". This procedure continues until some termination criterion (for example the number of maximum features to be used or admissible maximum classification error) is met. In this case, the goodness of almost all uncorrelated bands, but sorted as given by the SFFS algorithm, i.e., in terms of their flaw detection capability, will be evaluated for defect discrimination purposes. A schematic diagram of the overall band selection procedure is depicted in Figure 3.

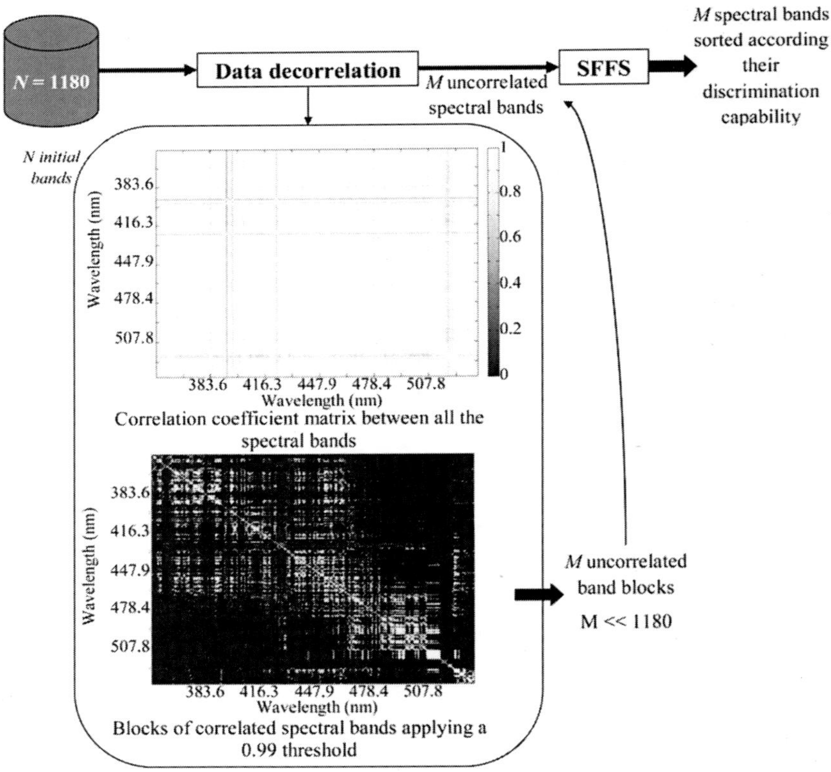

Figure 3. Block diagram of the spectral band selection and sorting procedure.

4. EVALUATION AND DISCUSSION OF EXPERIMENTAL AND FIELD TESTS

Analyses of different types of weld processes have been performed to study the influence of the spectral band selection on the resulting output line-to-continuum profiles. This issue was initially explored in a previous publication by our group [20], where an alternative technique was proposed, but line-to-continuum profiles were generated to allow comparison between both approaches. It seemed clear that the obtained results exhibited strong dependence on the chosen spectral band, what gave rise to the suitability of including a selection method in this regard.

Initial analyses were conducted on AISI-304 stainless steel plates, performing bead-on-plate seams with a TIG (tungsten inert gas) setup. The plasma optical radiation was collected by means of a 2 m optical fiber (50 μm core diameter) attached to an Ocean Optics USB2000 CCD-spectrometer with a spectral range from 195 to 535 nm. Argon was used as shielding gas, with a flow rate in normal operating conditions of 12 L/min. Three defects were provoked on two seams by simulating a perturbation on the gas flow, manually performing a shortening to 2 L/min during approximately 0.5 s. These seams are presented in Figure 4a,b, where the defects under analysis have been highlighted. The use of the SFFS algorithm for these two experimental tests provided the set of spectral bands listed in Table 1.

Table 1. Spectral bands selected and listed by SFFS and related to species participating in the plasma: experimental tests.

Spectral band (nm): order	Species	Spectral band (nm): order	Species	Spectral band (nm): order	Species
375.03: (1)	Cont.	357.93: (9)	Cr I	393.04: (17)	Cr I *
518.41: (2)	Cr I	396.48: (10)	Cr I	488.14: (18)	Ar II *
392.87: (3)	Fe I	403.33: (11)	Mn I	382.00: (19)	Fe I
481.08: (4)	Mn I *	396.64: (12)	Cr I	355.74: (20)	Ni I
434.75: (5)	Cr I	435.70: (13)	Cr I *	357.09: (21)	NI I *
480.62: (6)	Ar II	439.95: (14)	Fe I	423.00: (22)	Ar II *
487.84: (7)	Ar II	397.13: (15)	Cr I *	356.75: (23)	Ni I
360.63: (8)	Cr I	357.26: (16)	Cr I *		

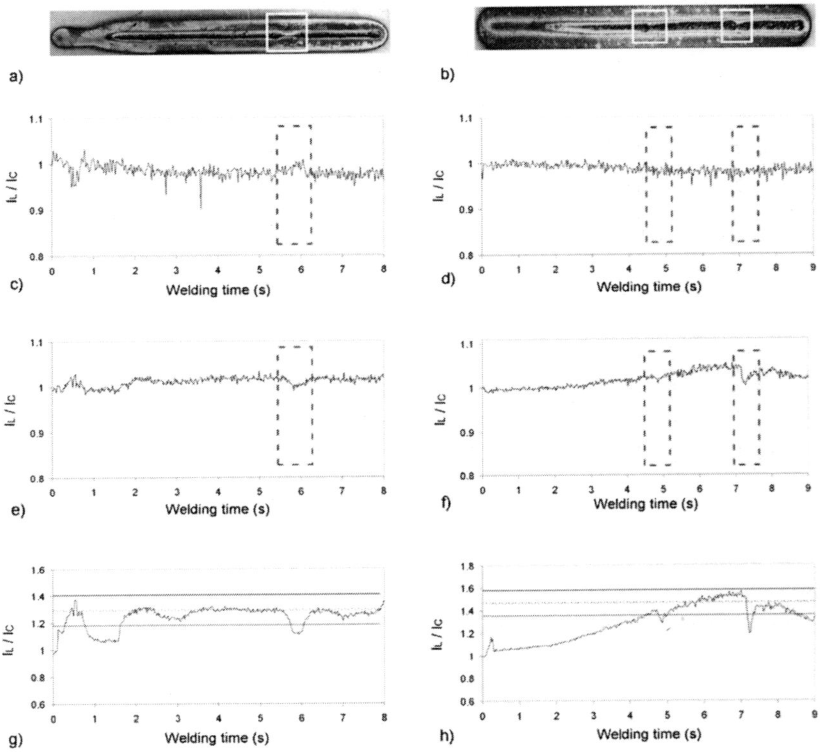

Figure 4. Detection of weld flaws by means of line-to-continuum method: a) and b) bead-on-plate seams with defects; c) and d) output profiles with spectral band 518.41 nm; e) and f) output profiles with spectral band 375.03 nm; g) and h) output profiles with spectral band 481.08 nm.

It can be observed that the selected emission lines have been also identified, and precisely the first one is not associated with an emission line but to continuum. * denotes that the spectral band is not centered on a particular emission line, but it is affected by its changes, i.e., it lies in the vicinity of that line. Figure 4 shows the discrepancies to be found in the line-to-continuum profiles depending on the spectral band selected. In these profiles I_L stands for the relative intensity of the chosen spectral band and I_C for the corresponding background radiation.

Figure 4c,d presents the results obtained with spectral band 518.41 nm, Figure 4e,f with 375.03 nm and Figure 4g,h with 481.08 nm. The latter offers the best results in terms of defect discrimination, given that the three flaws located at x ≈ 6 cm for seam n° 1 [Figure 4a], and at x ≈ 5 and 7 cm for seam n° 2 (Figure 4b) are associated with rapid perturbations on the profiles. On the contrary, Figure 4c,d does not show any correlation with these defects, indicating that the selected spectral band is not suitable in this case. By using the spectral band located at 375.03 nm the identification of the defects is again feasible, but the signal-to-noise ratio is clearly poorer in this case. It is worth mentioning that the dip at the beginning ofFigure 4g is due to the use of a lower welding current at the beginning of the process. As expected, a lower plasma electron temperature is associated with a lower welding current, given that the line-to-continuum profile is directly related to this parameter as suggested by Equation (3). The slope of Figure 4h was caused by a non-constant distance between the electrode tip and the plate provoked by the deformation of the plate by heat during previous seams. Again, the slope is associated with T_e, being the temperature higher as the distance between the electrode tip and the plate becomes smaller. These two situations can be also considered as defective, although the analysis via SFFS has been performed considering only the sections highlighted in Figure 4a,b as weld defects.

An example of automatic weld defect detection has been included in Figure 4g, using the approach proposed by Ancona*et al.* in [16]. In this case the reference signal has been generated considering the seam section between x ≈ 3.5 and x ≈ 5.5 cm as correct. The corresponding thresholds have been plotted by determining both the I_L/I_C mean and standard deviation values, and chosing $\alpha = 10$ [16]. It can be observed that not only the defect at x ≈ 6 cm, but also the lack of penetration at the beginning of the seam would be identified as defective, as the monitoring signal exceeds the calculated thresholds. The same analysis has been conducted in Figure 4h, where again the signal associated with the defects under analysis and the initial section of the seam exceeds the thresholds.

The results related to spectral bands 375.03 and 518.41 nm are worse than expected taking into account that their selection by the SFFS algorithm is performed earlier than the selection of the 481.08 nm spectral band. An initial analysis indicates that a possible explanation lies in the suppression of bands correlated to these ones that would offer better correlation to the quality of the seams. This occurs for example between spectral bands located at 375.03 and 446.07 nm. The latter gives rise to profiles similar to the ones depicted in Figure 4g,h, but this band is correlated to the former and, consequently, suppressed by the SFFS algorithm.

The plasma spectra for both experimental tests and spectral bands have been represented in Figure 5. It can be appreciated how the band chosen by SFFS does not belong to an emission line, while in 446.07 nm a significant peak is to be found.

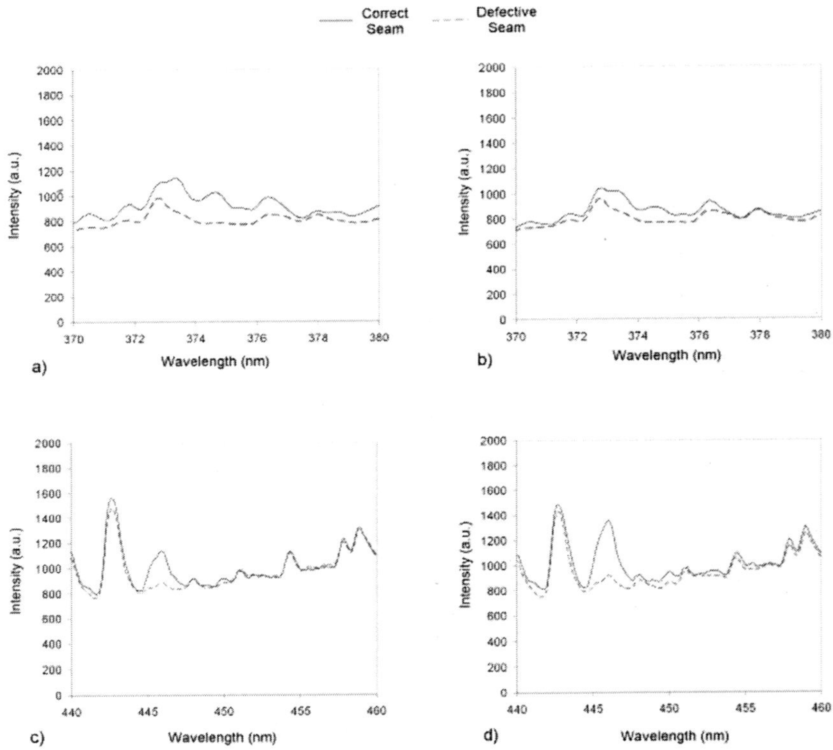

Figure 5. Comparison of spectra associated with correct and defective seams: a) and b) spectral band 375.03 for seams n° 1 and 2; c) and d) spectral band 446.07 for seams n° 1 and 2.

A feasible explanation for the uncertainty in the SFFS sorting criterion of the spectral bands is based on its assumptions about the probability distribution functions of the data. The employment of the SFFS algorithm implies a Gaussian or normal distribution of the classes. Correct welding could fit with this assumption but this is not necessarily the case of the defect class taking into the account the complexity of the physical phenomena occurring during the processes. An easy way to estimate whether the class distributions are Gaussian or not is by evaluating its third and fourth order statistics moments, namely skewness and kurtosis [29].

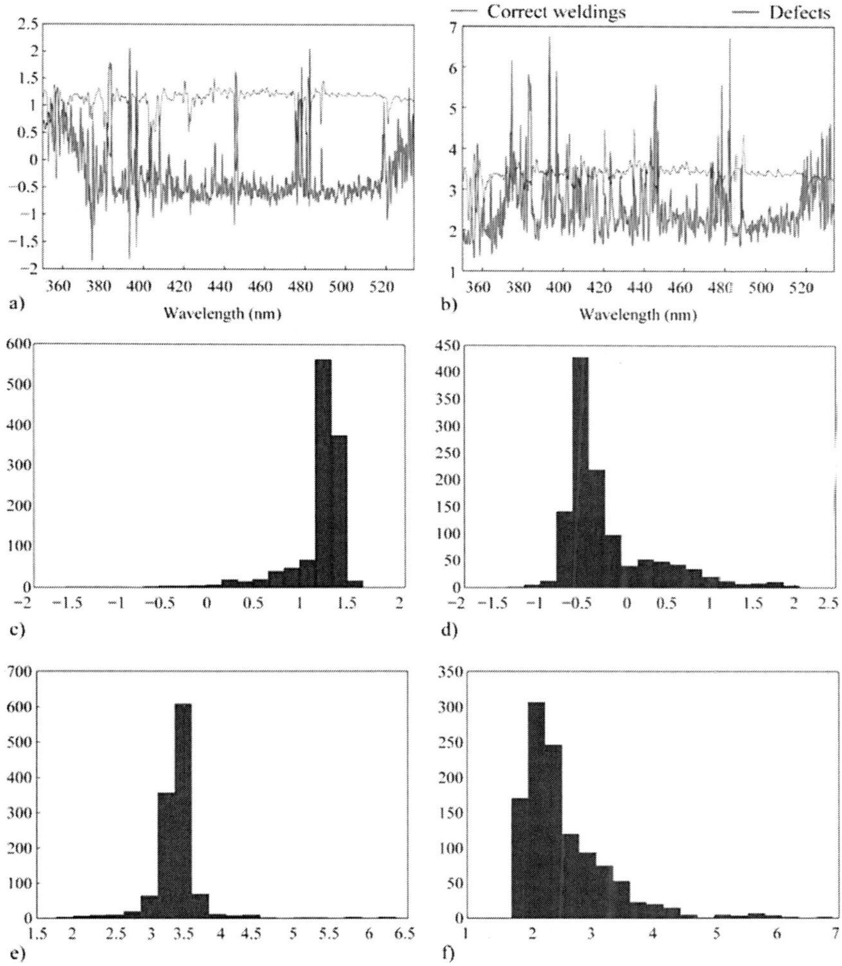

Figure 6. Modelling of the data distribution function of the classes: a) and b) skewness and kurtosis values per class and wavelength; c) histogram of skewness values of correct weldings d) histogram of skewness values of defects; e) and f) idem for kurtosis values.

Skewness is a measure of the asymmetry of the data around the class mean. Therefore the skewness of the normal distribution (as in any other perfectly symmetric distribution) is zero. Kurtosis is a measure of how outlier-prone a class distribution is and the kurtosis of the normal distribution is 3. Skewness and kurtosis values per class and wavelength are displayed in Figure 6a,b, respectively, while Figure 6c-f exhibits their corresponding histograms. Neither the skewness histogram pertaining to the correct welding class nor the one pertaining to defects are centered around zero, which allows us to conclude that none of the classes have a normal distribution. In spite of this fact, SFFS provides the selection of some spectral bands that allow defect detection such as the one in 481.08 nm and, as a consequence, a two step methodology is proposed. First the SFFS algorithm is applied and the line-to-continuum profiles are obtained for the selected wavelengths. Then the signal-to-noise ratio of these profiles is evaluated and the one with the highest value is selected as the monitoring signal.

The reordering of the spectral bands in terms of the S/N^* (Defect Sensitivity) parameter is presented in Table 2. A detailed study of the resulting line-to-continuum profiles was conducted to elucidate a suitable method to quantify their sensitivity to detect the weld flaws. This S/N^* is calculated as follows: first, the signal-to-noise ratio over a section free of defects is estimated (S/N in Table 2), being in this case the seam segment between welding times 3 and 5 s for seam n° 1. Afterwards, the defect detection sensitivity is considered as the difference between the I_L/I_C value at the defect (x ≈ 6 s) and x = 5.5, where the signal indicates a correct seam. Finally, this two values are multiplied to generate the S/N^* parameter. This new ordering of the spectral bands is in good agreement with the results shown in Figure 4, as the spectral bands corresponding to Figure 4c,e,g is classified with this new method in 19th, 16th and 4th positions.

The proposed technique has been also checked with data obtained during field tests of a tube-to-tubesheet orbital TIG welding process. The material to be welded was INCONEL 690 tubes, and the tests were developed on a welding coupon simulating the conditions of a steam generator for a nuclear power station. The process is described in detail in [30].

As the material and process parameters were completely different from the previously discussed experimental tests, a new set of spectral bands had to be generated with SFFS. The new bands are displayed in Table 3, and Figure 7 depicts the chosen output profiles for a tube with a defective section. The S/N^* parameter has been included in the Table, being in this case estimated with the field test shown in Figure 7. The flaw was provoked by a perturbation on the gas flow rate by the operator, what gave rise to the defective section highlighted in Figure 7a.

Table 2. Spectral bands selected and listed by SFFS and reordered by S/N Defect Sensitivity.

Spectral band (nm): SFFS-order	Species	S/N	S/N * Defect Sensitivity (order)
403.33: (11)	Mn I	55.69	18.93 (1)
357.09: (21)	NI I *	66.91	18.73 (2)
357.26: (16)	Cr I *	74.67	16.43 (3)
481.08: (4)	Mn I *	80.5	15.22 (4)
446 (**)	Mn I	54.69	12.03 (5)
382.00: (19)	Fe I	101.71	10.17 (6)
393.04: (17)	Cr I *	19.38	9.69 (7)
356.75: (23)	Ni I	30.89	9.27 (8)
357.93: (9)	Cr I	45.94	9.19 (9)
396.64: (12)	Cr I	22.19	8.87 (10)
360.63: (8)	Cr I	58.07	8.71 (11)
392.87: (3)	Fe I	24.23	7.95 (12)
396.48: (10)	Cr I	25.16	7.55 (13)
397.13: (15)	Cr I *	37.41	7.48 (14)
423.00: (22)	Ar II *	71.84	7.18 (15)
375.03: (1)	Cont.	197.81	5.14 (16)
355.74: (20)	Ni I	36.6	4.39 (17)
487.84: (7)	Ar II	33.51	2.01 (18)
518.41: (2)	Cr I	85.16	1.79 (19)
435.70: (13)	Cr I *	89.29	1.79 (20)
480.62: (6)	Ar II	26.07	1.56 (21)
488.14: (18)	Ar II *	37.68	1.13 (22)
434.75: (5)	Cr I	27.55	0 (23)
439.95: (14)	Fe I	41.51	0 (24)

It can be appreciated that the defective section is associated with perturbations on the I_L/I_C profiles, while the seam labelled as correct is more stable. The perturbation on the protection gas flow is reflected on the welding plasma, and consequently also on the intensity of the different emission lines versus the continuum radiation. The operator catalogued the seam as defective after visual inspection, given the irregularities to be observed on the surface, as well as the associated bead color. Further analyses by means of destructive or non-destructive evaluation techniques could have helped to improve this classification, but in this case visual inspection was considered to be adequate.

According to the ISO 6520-1:1998 standard this defect could be labelled as an irregular surface (excessive surface roughness). Four different spectral bands have been considered in this case: 404.14, 422.84, 480.52 and 423.64 nm. It is precisely the latter, associated with Fe I, the one which exhibits worst results in terms of sensitivity to defects, as all the profiles have been represented with the same vertical scale. It is important highlighting that the defective section could be identified, at least partially, in the four cases, but the profile in Figure 7c offers the best sensitivity: in terms of S/N^* the spectral bands are classified as 7th (404.14 nm), 2nd (422.84 nm), 8th (480.52 nm) and 10th (423.64 nm). The I_L/I_C thresholds have been calculated again for Figure 7c, using in this case $\alpha = 3$.

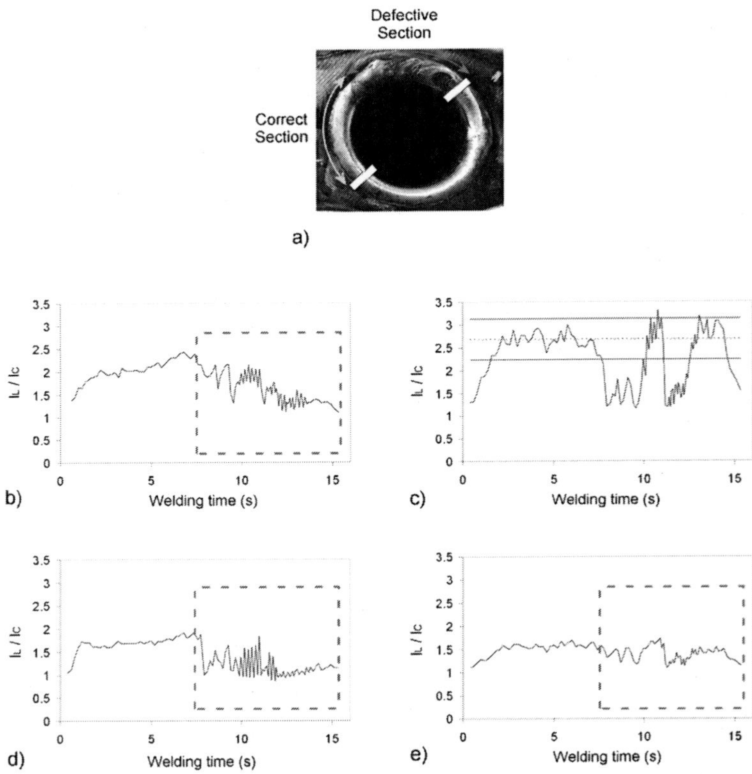

Figure 7. Field test tube with defective section: a) welded tube; b) profile for spectral band 404.14 nm; c) profile for spectral band 422.84 nm; d) profile for spectral band 480.52 nm; e) profile for spectral band 423.64 nm.

Table 3. Spectral bands selected and listed by SFFS and related to species participating in the plasma: field tests.

Spectral band (nm): (SFFS-order)	Species	S/N	S/N * (order)
516.79 (12)	Fe I	29.1	41.03 (1)
422.84 (3)	Ar II	16.1	24.30 (2)
415.95 (2)	Ar I	20.96	20.54 (3)
402.83 (6)	Mn I *	30.68	19.68 (4)
426.35 (5)	Cr I *	28.42	15.92 (5)
407.39 (9)	Cont.	42.69	14.94 (6)
404.14 (1)	Mn I *	13.91	14.47 (7)
480.52 (7)	Ar II	20.24	14.37 (8)
443.54 (11)	Fe I	28.57	14.00 (9)
423.64 (4)	Fe I	26.1	11.48 (10)
478.28 (8)	Mn I	26.28	11.30 (11)
432.28 (14)	Cont.	36.9	7.38 (12)
427.30 (10)	Cr I	52.77	3.69 (13)
520.05 (13)	Cr I	19.1	1.53 (14)

A different defect is presented in Figure 8a, where a porosity (highlighted in red) was created due to the application of a gel to the tube-to-tubesheet interface. It is interesting to note that, although only the porosity appears as visible, the gel was applied trough the whole interface, what probably provoked deviations from the standard quality requirements. This can be the explanation to the fast perturbations that appear in the profiles of Figure 8b-e. Again, Figure 8c, where the I_L/I_Cthresholds have been plotted, offers the best identification of the porosity and the higher defect sensitivity. It is interesting to note that in this case the classification provided by *S/N** seems valid for the first three bands, but the fourth (Figure 8e) exhibits a better response than Figure 8b, for example. This can be explained because the *S/N** analysis was performed with the field test associated with Figure 7. In this case the generalization of these classifications to both seams would not be valid, and a separate study should be conducted.

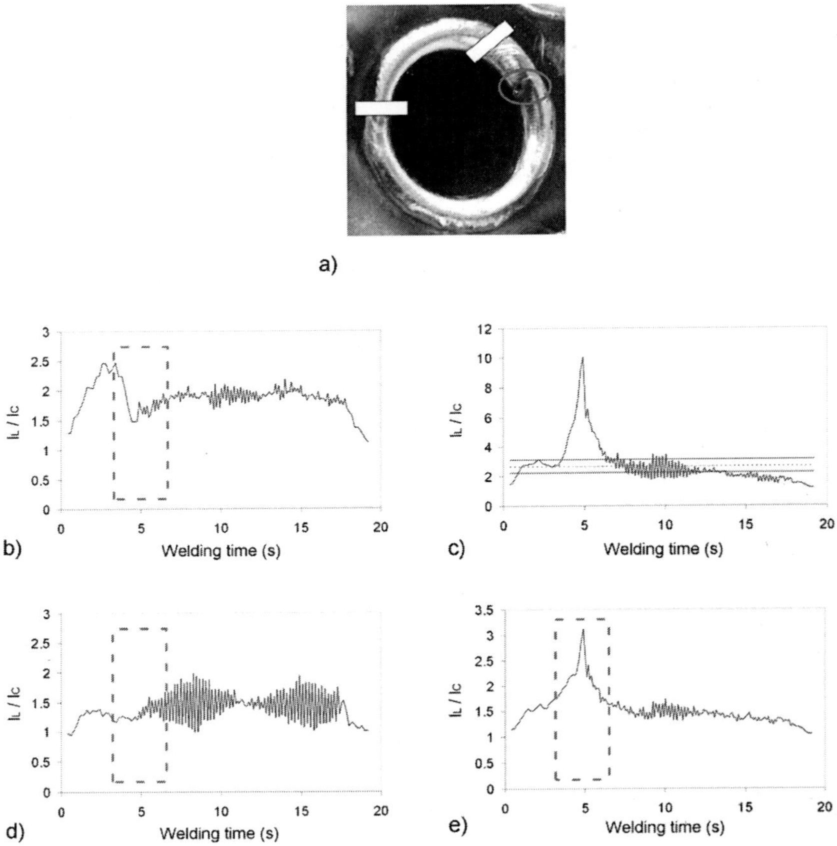

Figure 8. Field test tube with porosity: a) welded tube; b) profile for spectral band 404.14 nm; c) profile for spectral band 422.84 nm; d) profile for spectral band 480.52 nm; e) profile for spectral band 423.64 nm.

5. CONCLUSIONS

A spectroscopic approach based on feature selection and the line-to-continuum method for on-line detection of welding defects is proposed and experimentally validated in this paper. Compared with the determination of the plasma electron temperature, the main advantage of the line-to-continuum method is its computational efficiency, since it does not imply the identification of the atomic emission lines. However, its defect detection capability highly depends on the chosen emission lines. Therefore, an initial stage is required to determine those spectral bands from the welding plasma that best discriminate among correct welding and the appearance of defects. In a previous work [22] the SFFS algorithm was employed to reduce spectral dimensionality of data from an embedded fiber-sensor for on-line welding diagnostics. Selection criterion of the spectral bands was based on the maximization of the distance between the different flaws and correct welds. This same approach has been applied here to get rid of the uncertainty regarding the selection of the optimum spectral bands. SFFS is firstly utilized to identify the proper lines for defect detection. Then, the monitoring signal consists of the line-to-continuum profiles of those lines. Experimental results show a strong correlation between the appearance of flaws and abrupt changes in the monitoring signal.

The conducted studies, which include an extended set of defects with different target materials, have demonstrated the suitability of the methodology for on-line analysis. However, they have also brought up that some uncertainty still remains regarding the SFFS sorting criterion of the spectral bands. As shown, it occurs that some of the spectral bands that are selected later on by SFFS have proved more successful in defect-detection. This is due to the fact that the wavelength selection criterion, the Bhatacharyya distance, assumes a Gaussian distribution of the classes and neither correct weldings nor the defect class fit with this assumption. As a consequence, a further comparative study on the capability of the selected wavelengths for flaw detection has been performed, being the signal-to-noise ratio of the line-to-continuum profiles the comparison criterion. Regarding the use of the S/N^* parameter to perform this fine classification of the spectral bands, it would be interesting to define a more general approach able to quantify the sensitivity to the appearance of different defects. Current studies are on going focused on this objective and also on the classification of the defects in order to actuate on the precise welding parameter to try to prevent each defect from happening. It should be also investigated if a set of different spectral bands is needed to accomplish discrimination between different defects, or an approach based in a single band is feasible, even to be used for different materials and processes.

ACKNOWLEDGMENTS

This work has been co-supported by the project TEC2007-67987-C02-01. Authors also want to thank J.J. Valdiande for his valuable support during the experimental welding tests.

REFERENCES AND NOTES

1. Eagar, T.W. Physics of Processes. In *Advanced Joining Technologies*; Chapman and Hall: London, UK, 1990.
2. Wu, C.S.; Ushio, M.; Tanaka, M. Analysis of the TIG welding arc behaviour. *Computat. Mater. Sci.* **1997**, *7*, 308–314.
3. Haidar, J. A theoretical model for gas metal arc welding and gas tungsten arc welding. I. *J. Appl. Phys. (USA)* **1998**, *84*, 3518–3529.
4. Gornushkin, I.B.; Kazakov, A.Y.; Omenetto, N.; Smith, B.W.; Winefordner, J.D. Experimental verification of a radiative model of laser-induced plasma expanding into vacuum. *Spectrochim. Acta B* **2005**, *60*, 215–230.
5. Halmshaw, R. *Introduction to the non-destructive testing of welded joints*; Cambridge: Abington, UK, 2006.
6. Li, L.; Brookfield, D.J.; Steen, W.M. Plasma charge sensor for in-process, non-contact monitoring of the laser welding process. *Meas. Sci. Technol.* **1996**, *7*, 615–626.
7. Lu, W.; Zhang, Y.M.; Emmerson, J. Sensing of weld pool surface using non-transferred plasma charge sensor. *Meas. Sci. Technol.* **2004**, *15*, 991–999.
8. Gu, H.; Duley, W.W. Statistical approach to acoustic monitoring of laser welding. *J. Phys. D* **1996**, *29*, 556–560.
9. Farson, D.F.; Kim, K.R. Generation of optical and acoustic emissions in laser weld plumes. *J. Appl. Phys.* **1999**, *85*, 1329–1336.
10. Zhang, G.J.; Yan, Z.H.; Wu, L. Visual sensing of weld pool in variable polarity TIG welding of aluminium alloy. *Trans. Nonferrous Met. Soc. China* **1997**, *16*, 522–526.
11. Kovacevic, R.; Zhang, Y.M.; Ruan, S. Sensing and control of weld pool geometry for automated GTA welding. *J. Eng. Ind. Trans. ASME* **1995**, *117*, 210–222.
12. Gu, H.; Duley, W.W. Possible diagnostic signal for monitoring CO2 laser welding of aluminum alloy sheets. *Proc. SPIE* **1995**, *2374*, 208–214.
13. Bardin, F.; Cobo, A.; Lopez-Higuera, J.M.; Collin, O.; Aubry, P.; Dubois, T.; Hogstrom, M.; Nylen, P.; Jonsson, P.; Jones, J.D.C.; Hand, D.P. Optical techniques for real-time penetration monitoring for laser welding. *Appl. Opt.* **2005**,*44*, 3869–3876.

14. Sforza, P.; de Blasiis, D. On-line optical monitoring system for arc welding. *NDT E Int.* **2002**, *35*, 37–43.

15. Ancona, A.; Spagnolo, V.; Lugara, P.M.; Ferrara, M. Optical sensor for real-time monitoring of CO_2 laser welding process. *Appl. Opt.* **2001**, *40*, 6019–6025.

16. Ancona, A.; Maggipinto, T.; Spagnolo, V.; Ferrara, M.; Lugara, P.M. Optical sensor for real time weld defects detection. *Proc. SPIE* **2002**, *4669*, 217–226.

17. Mirapeix, J.; Cobo, A.; Conde, O.M.; Jaúregui, C.; López-Higuera, J.M. Fast algorithm for spectral processing with application to on-line welding quality assurance. *Meas. Sci. Technol.* **2006**, *17*, 2623–2629.

18. Sibillano, T.; Ancona, A.; Berardi, V.; Schingaro, E.; Parente, P.; Lugara, P.M. Correlation spectroscopy as a tool for detecting losses of ligand elements in laser welding of aluminium alloys. *Opt. Lasers Eng.* **2006**, *44*, 1324–1335.

19. Mirapeix, J.; Cobo, A.; González, D.A.; Lopez-Higuera, J.M. Plasma spectroscopy analysis technique based on optimization algorithms and spectral synthesis. *Opt. Express* **2007**, *15*, 1884–1897.

20. Mirapeix, J.; Cobo, A.; Fernandez, S.; Cardoso, R.; Lopez-Higuera, J.M. Spectroscopic analysis of the plasma continuum radiation for on-line arc-welding defect detection. *J. Phys. D* **2006**, *41*, 135202–135210.

21. Mirapeix, J.; García-Allende, P.B.; Cobo, A.; Conde, O.M.; López-Higuera, J.M. Real-time arc-welding defect detection and classification with Principal Component Analysis and Artificial Neural Networks. *NDT E Int.* **2007**, *40*, 315–323.

22. Garcia-Allende, P.B.; Mirapeix, J.; Conde, O.M.; Cobo, A.; Lopez-Higuera, J.M. Arc-welding spectroscopic monitoring based on feature selection and neural networks. *Sensors* **2008**, *8*, 6496–6506.

23. Griem, H.R. *Principles of Plasma Spectroscopy*; Cambridge University Press: New York, NY, USA, 1997.

24. Marotta, A. Determination of axial thermal plasma temperatures without Abel inversion. *J. Phys. D* **1993**, *27*, 268–272.

25. Bastiaans, G.J.; Mangold, R.A. The calculation of electron density and temperature in Ar spectroscopic plasmas from continuum and line spectra. Spectrochim. *Acta A* **1985**, *40B*, 885–892.

26. Ferri, F.; Pudil, P.; Hatef, M.; Kittler, J. Comparative study of techniques for large-scale feature selection. In *Pattern Recognition in Practice IV: Multiple Paradigms, Comparative Studies, and Hybrid Systems*; Gelsema, E.S., Kananl, L.N., Eds.; Elsevier Science: New York, NY, USA, 1994; pp. 403–413.

27. Gomez-Chova, L.; Calpe, J.; Camps-Valls, G.; Martin, J.D.; Soria, E.; Vila, J.; Alonso-Chorda, L.; Moreno, J. Feature selection of hyperspectral data through local correlation and SFFS for crop classification. *IEEE Int. Geosci. Remote Sens. Symp. Proc.* **2003**, *1*, 555–557.

28. Deronde, D.; Kempeneers, P.; Forster, R.M. Imaging spectroscopy as a tool to study sediment characteristics on a tidal sandbank in the Westerschelde. *Estuar. Coast. Shelf Sci.* **2006**, *69*, 580–590.

29. Holub, O.; Ferreira, S.T. Quantitative histogram analysis of images. *Comput. Phys. Commun.* **2006**, *175*, 620–623.

30. Cobo, A.; Mirapeix, J.; Linares, F.; Piney, J.A.; Solana, D.; Lopez-Higuera, J.M. Spectroscopic Sensor System for Quality Assurance of the Tube-To-Tubesheet Welding Process in Nuclear Steam Generators. *IEEE Sensors J.* **2007**, *7*, 1219–1224.

4

CHAPTER

A Real-Time Spectroscopic Sensor for Monitoring Laser Welding Processes

Teresa Sibillano [1,*], Antonio Ancona [1], Vincenzo Berardi [2] and Pietro Mario Lugarà [1,2]

[1]CNR-INFM Regional Laboratory LIT[3], via Amendola 173, Bari, Italy

[2]Dipartimento Interateneo di Fisica, Università e Politecnico di Bari, Via E. Orabona 4, I-70126 Bari, Italy

ABSTRACT

In this paper we report on the development of a sensor for real time monitoring of laser welding processes based on spectroscopic techniques. The system is based on the acquisition of the optical spectra emitted from the laser generated plasma plume and their use to implement an on-line algorithm for both the calculation of the plasma electron temperature and the analysis of the correlations between selected spectral lines. The sensor has been patented and it is currently available on the market.

Keywords: Optical sensor; laser welding; plasma spectroscopy

1. INTRODUCTION

Traditional off-line inspection of welded joints is expensive and reduces productivity, and the lack of effective on-line controls in laser machining is one of the main obstacles for the full implementation of laser welding technologies in industrial applications.

Several solutions have been proposed in recent years for the development of automated on-line laser welding monitoring sensors. Spectroscopic investigation of the plasma optical emission provides a number of potential advantages for a detailed analysis of defects as a function of the laser operation parameters and the material properties.

In this work an overview will be given of the recently developed optical-based monitoring systems for laser welding processes. Then, we will focus on our last experimental results on the development of an optical sensor, based on plasma spectroscopy, especially conceived for real-time control and optimization of the welding processes using a CO_2 laser source.

2. OPTICAL SENSING FOR LASER WELDING

In laser welding the laser-metal interaction is usually associated with the ejection of material from the interaction area. The ejected material contains excited atoms and ions and it is commonly named plume. The material moves through the incident beam and is thus further heated to temperatures exceeding the vaporization temperature. Under certain conditions, the overall effect is to produce a rapid increase in the level of ionization within the plume with the formation of a plasma. Because the plasma is created only when vaporization occurs, its presence during laser welding may provide useful information about the welding conditions.

Several signals coming from the plasma can be used to yield information on the possible presence of defects during the process [1]. Among the many possible techniques for the development of optical sensors, the most effective ones are those based on the measurement of the spatially integrated optical intensity by one or more photodiodes as well as the spectroscopic analysis of the UV/VIS emission [2,3]. Such kind of optical sensors based on the plasma plume optical emission have been used successfully in industrial environments for the detection of several types of welding defects.

Several works have reported on the development of optical sensors based on photodiodes as a tool for the detection of the light intensity from the plasma produced during the process. Gaztweiler *et al.* [4] monitored the keyhole plasma by using an array of photodiodes collecting light at different viewing angles with respect to the beam axis. By this arrangement each detector monitors a different region of the plasma. The overall plasma intensity distribution inside the keyhole was then estimated by combining the signals. Such a sensor was reliable in monitoring the penetration depth on thick samples of steel sheets as well as in obtaining information on the bead shape.

Park *et al.* [5,6] monitored both the bead shape and the full penetration by a photodiode-based acquisition of the UV emission from plasma and the IR from spatters. A simultaneous measurement of the spectral line intensities and of the IR radiation from the weld pool gave a possible correlation between the plasma characteristics and the size and temperature of the weld pool. In this way the authors show that simultaneous measurements of the spectral line intensities from the plasma emission, together with IR emission from the weld pool, provide a way to relate changes in plasma characteristics to the size and temperature of the weld pool itself. The limit of this approach is the difficulty to separate the IR emission coming from the weld pool from the IR signal coming from the plasma.

Peters *et al.* [7,8] reported a non-intrusive optical sensing technique for Nd:YAG laser welding based on the simultaneous detection of the light radiated by the plasma plume above the welding surface and of the light back propagating in the cladding of the laser beam delivery fiber (cladding power monitor, CPM). The cladding was coupled to the core of the monitor fiber and then delivered to a UV/VIS photodiode. The system was used to demonstrate the correlation between the optical signals of the plume and laser welding faults. The same group has recently reported [9,10] on a real-time focus control during Nd:YAG laser welding.

Bardin *et al.* [11,12] described the design of a closed-loop system that monitors the focal position to ensure full penetration during Nd:YAG laser welding processes. The focus position monitoring system was based on the chromatic aberration of the focusing optics: the results obtained showed that the spectral analysis of the light emitted from the weld pool detected by three different photodiodes gave information on the focal error.

To optimize the process parameters, other authors have investigated the stability of photodiode signals by monitoring fluctuations. A closed loop control system has been developed by Bagger and Olsen [13] to control the laser power by observing the light emission from the root-side of the sheet. The control system is able to adjust the laser power to keep an even root seams if the welding speed or the thickness are changed. The main drawback of such a system is that it is not suitable for lap welding with a thickness greater than 1.25 mm.

2.1. Plasma spectroscopy

In laser welding it is well known that a strong plasma optical emission is observable right above the keyhole and can be easily collected by using optical fibers [14]. As a consequence, plasma plume optical spectroscopy is a very promising technique for realizing a reliable on-line monitoring of the quality of welded joints and in general for keeping under control the welding process. The spectroscopic approach has been easily extended to arc welding by several research groups [15,16], improving the performance of the arc-welding monitoring systems.

Plasma optical spectra are characterized by the presence of emission lines coming both from the excited atoms and from the ions produced during the laser-surface interaction. A careful spectroscopic characterization of such emission lines allows to determine the chemical composition and the dynamics of interaction of the different chemical species inside the plume.

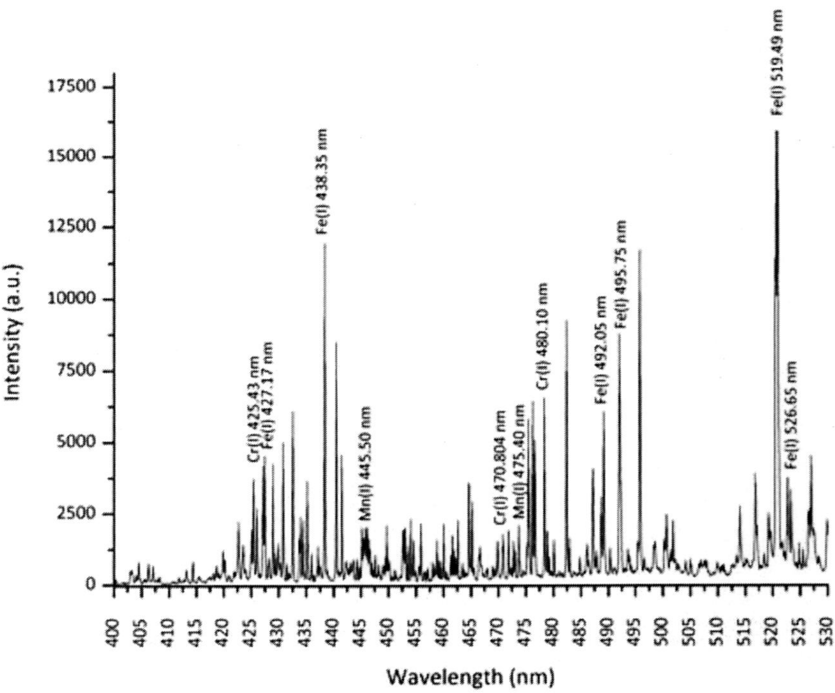

Figure 1. Plasma optical emission in the range 400-530 nm with a resolution of 0.07 nm, obtained during CO_2 laser welding of AISI 304 stainless steel.

The measurement of the plasma electron temperature as well as the analysis of the plasma optical spectra by using the Covariance Mapping Technique have been the subject of several papers published by our research group over the last few years [17-20]. The final objective was to combine the two above mentioned techniques to develop an optical sensor for real-time defect recognition during industrial laser processes.

The plasma electron temperature can be obtained from the measurement of the relative intensity of a set of spectral lines free from self-absorption and by the application of the Boltzamnn plot method. The intensity I_{mn} of a plasma emission line associated with the decay between levels E_m and E_n is related to the energy of the emitted photon, hc/λ_{mn}, the transition probability A_{mn}, and the population of the exited state N_m by the following equation:

$$I_{mn} = N_m A_{mn} hc / \lambda_{mn} \tag{1}$$

Assuming Boltzmann statistics, N_m can be expressed as:

$$N_m = (N / Z) g_m \exp(-E_m / kT) \tag{2}$$

where N is the total number of states, g_m is the degeneracy and Z is the partition function. From Eqs. 1 and 2 we can obtain:

$$\ln\left(\frac{I_{mn}\lambda_{mn}}{A_{mn}g_m}\right) = \ln\left(\frac{Nhc}{Z}\right) - \frac{E_m}{kT_e} \tag{3}$$

It is evident that Equation (2) shows a linear dependence of the left side of the equation from the level energy E_m. The Boltzman plot method consists then in plotting Equation (2) for several spectral emission lines belonging to the same chemical species and perform a linear fit. As evident from Equation (2), the electron temperature T_e is immediately inferred from the slope of the linear fit.

The electron temperature can be also estimated by use of the intensity ratio of just a pair of emission lines, labeled (1) and (2) in the following equation, among those selected for the Boltzmann plot:

$$\frac{I(1)}{I(2)} = \frac{A(1)g_m(1)\lambda(2)}{A(2)g_m(2)\lambda(1)} \exp\left[-\frac{E_m(1) - E_m(2)}{kT_e}\right] \tag{4}$$

Extracting T_e from Eq.4:

$$T_e = \frac{E_m(2) - E_m(1)}{k \ln\left[\frac{I(1)A(2)g_m(2)\lambda(1)}{I(2)A(1)g_m(1)\lambda(2)}\right]} \tag{5}$$

Since the emission line parameters are well known, this method is particularly useful because it does not require too many calculations and can easily be implemented in real-time measurements.

The principle of Covariance Mapping Technique (CMT) is based on the assumption that the spectrum under investigation can be considered as a sampling function for the signal coming from the detector. Under this hypothesis, if $x_k(\lambda_i)$ is the signal recorded at wavelength λ_i of the k-th spectrum, the covariance matrix of the series of N spectra is given by:

$$C_{ij} = \frac{1}{N}\sum_{k=1}^{N} x_k(\lambda_i)x_k(\lambda_j) - \left[\frac{1}{N}\sum_{k=1}^{N} x_k(\lambda_i)\right]\left[\frac{1}{N}\sum_{k=1}^{N} x_k(\lambda_j)\right] \tag{6}$$

The covariance matrix is clearly symmetric upon exchange of *i* and *j* and in our analysis, we use the normalized form of the matrix, defined by:

$$m_{ij} = \frac{C_{ij}}{(C_{ii} \cdot C_{jj})^{1/2}} \tag{7}$$

The value of m_{ij} lies in the range between -1 and 1, and the diagonal elements m_{ii} are equal to 1. $m_{ij} = 1$ indicates correlated pairs of chemical species, whereas $m_{ij} = -1$ indicates anti-correlated lines of the spectrum. In this condition each peak of the spectrum can be viewed as a region of several points which can or cannot be correlated with other regions of the spectrum itself.

Correlated or anti-correlated regions are the result of correlated or anti-correlated light emissions which are distinguishing features of the chemical species present in the plasma plume. The main feature of the CMT resides in its reliability in extracting the information in case of very weak lines, provided that change in intensity can be statistically correlated to the variation of the intensity of some other lines. The CMT is thus particularly powerful to evidence the contribution of weak spectral signals which carry information but are characterized by a small value of the signal/noise ratios.

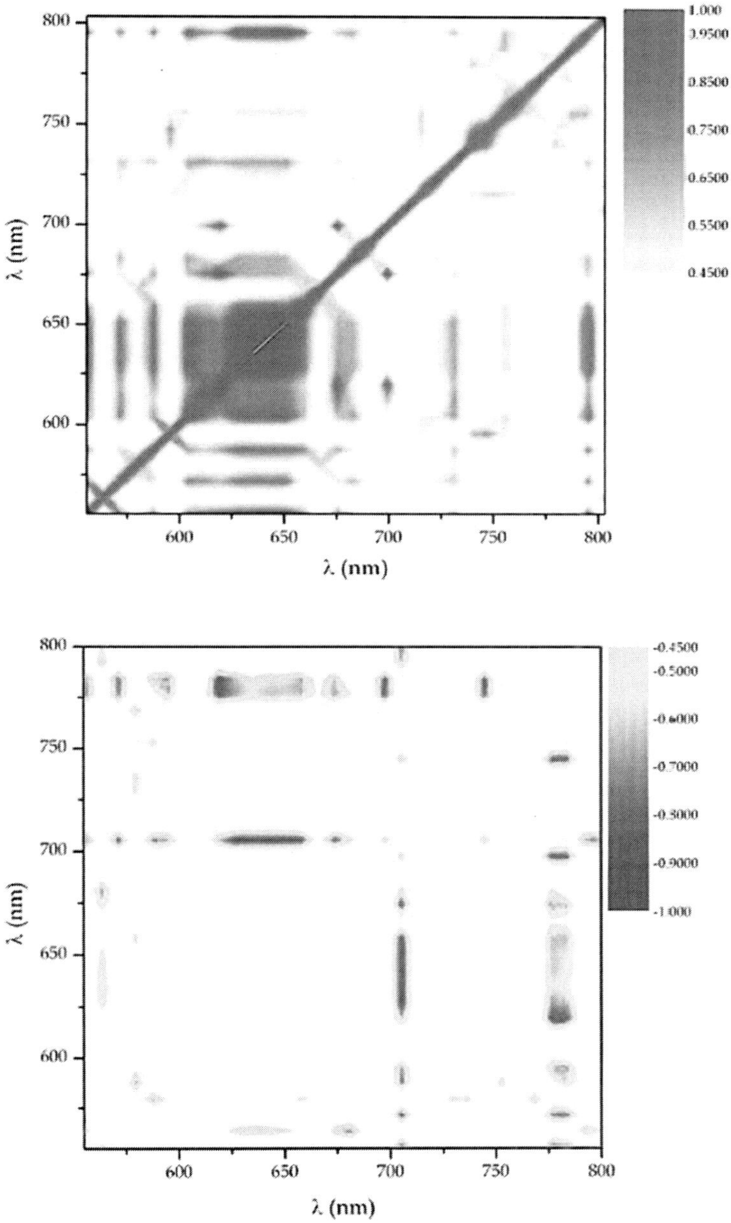

Figure 2. Correlation (top) and anti-correlation (bottom) maps for Al-Mg aluminium alloys.

In Figure 2 two samples of the correlation and anti correlation maps in the case of laser welding of Al-Mg Aluminium alloy are shown. A positive correlation value between the intensity of emission coming from two different species, identified by the emission wavelength and seen as reddish zones in the upper map, indicates that the evolution of their concentration in the plasma plume is the consequence of a process that has similar features for both the investigated species. On the other hand, a negative correlation means that the two species are formed by competing processes. In case of non-correlation, the species evolve through unrelated mechanisms and no useful information can be obtained. In our analysis, only the values of m_{ij} corresponding to a confidence level above 95% were considered. The threshold value fulfilling this criterion depends only on the number of spectra used in the covariance matrix evaluation.

3. EXPERIMENTAL PROCEDURE AND RESULTS

Several welding trials have been carried out in recent years to check whether the electron temperature and the correlation coefficient signals can be a valuable source of information for the development of a real-time monitoring sensor. The trails were conducted by using a high power CO_2 laser with a maximum output power of 2.5 kW. The plasma optical emission was collected by a quartz collimator of 6 mm focal length, and the light then transmitted to a PC interfaced miniaturized spectrometer. The spectral range investigated was 390-800 nm with a spectral resolution of 0.3 nm. The final result of our tests is an optical sensor that embeds a fiber-coupled miniature spectrometer with a dynamic spectral range from 390 nm to 580 nm and a resolution of 0.3 nm, which performs a real-time data acquisition and spectral analysis integrating the information given by the two spectroscopic techniques previously described.

The electron temperature was monitored for different chemical species inside the plasma plume, chosen according to the metal which is being welded finding a clear correlation between the temperature mean value, its standard deviation and the quality of the welded joints. This information was used to optimize the welding parameters. We also showed that local perturbations and/or variations of the temperature value are related to the occurrence of weld defects. The defects most sensitive for this technique are: i) lack of penetration; ii) weld disruptions; iii) crater formation and also iv) seam oxidation.

We created a defect detection algorithm by setting a reference value and two adjustable error thresholds for the electron temperature. The reference signal was computed during a self-learning procedure based on preliminary acquisition of sound welds performed in similar operating conditions. The upper and lower error thresholds were defined by adding or subtracting an adjustable fraction of the average standard deviation of sample signals, so setting the desired sensitivity of the monitoring system.

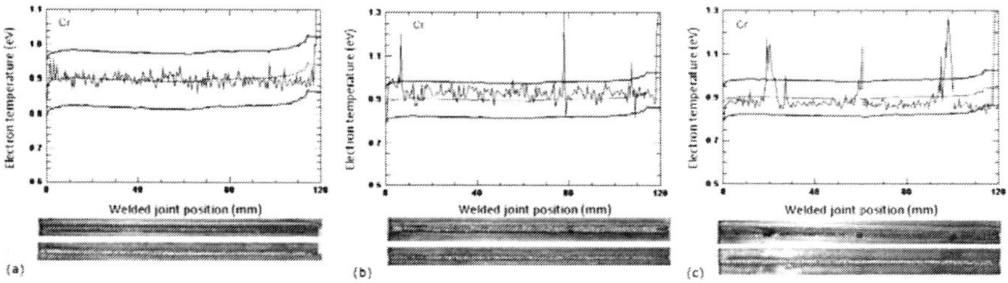

Figure 3. Electron temperature signals along the welded joints in presence of (a) a defect free weld (b) laser power variations and (b) crater formation.

Figure 3 shows the electron temperature signals correlated to the welded joints position in three different welding conditions; in (a) the electron temperature for defect-free weld is shown. It is evident that the electron temperature value is always in between the chosen thresholds. In (b) and (c), on the other hand, it is visible a clear increase of the electron temperature value for variable laser power and in presence of craters in three different points of the joints.

For what concerns the second technique used in the sensor, we have successfully used CMT to relate the dynamics of the chemical species present in the plasma plume to the characteristics of the welded joint. In case of laser welding of Al-Mg alloys, we have demonstrated that when there is either a collapse of the keyhole, due to the incorrect setting of the welding speed, or a loss of Magnesium, due to an inappropriate choice of the laser power, there is a marked change in the correlation signal between Al and Mg. The signal can flip from positive to negative correlation or viceversa [19, 20]. Needless to say the collapse of the keyhole causes always a defect in the welded joint.

Our findings have been confirmed by EDX analysis of the metallographic cross section of the allegedly defective welded joints. Moreover, we were able to correlate the influence of the welding speed on the loss of alloying elements by monitoring the correlation/anti-correlation between environmental gases (N and O) and Mg, which is a ligand of AA5083. As shown in Figure 4, we observed an high anti-correlation between Mg and

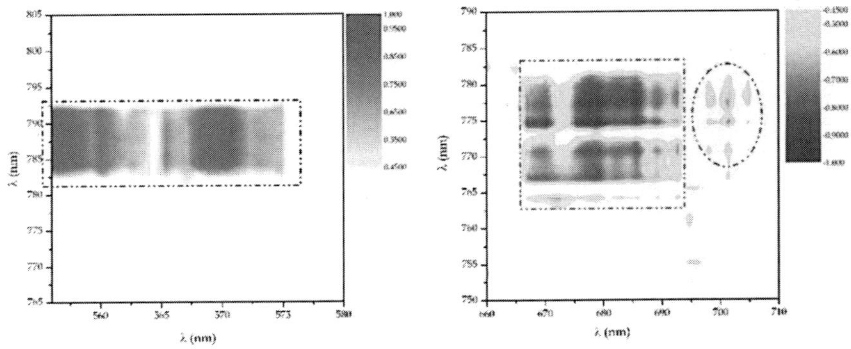

Figure 4. Correlation and anti-correlation map for Mg and Al for effective and ineffective shielding gas condition.

Al, in cases of ineffective shielding gas, consistently with the post-process EDX analyses performed on the welded joints and an high correlation between these two species for effective gas shielding conditions.

The measure of the correlation (anti-correlation) signal for a selected couple of lines allows to overcome the main drawback for the use of CMT in real-time sensors, that is the long acquisition and computation time requested for building the Covariance Map of the whole spectrum. By selecting only pairs of lines only a few number of the Covariance Matrix have to be evaluated paving the road towards the use of such a technique for real-time sensors [20].

Field tests have been carried out on an industrial production line of laser welded stainless steel tubes. The laser-metal interaction zone was shielded by argon flux. In this case Mn(I), Fe(I) and Cr(I) spectral lines were used for the acquisition of the electron temperature as well as for the correlation signals. The tested sensor was able of monitoring simultaneously the temporal evolution of the plasma plume and of the correlation coefficient during the welding process. For the covariance analyses, we used a sliding window of N=10 consecutive spectra. Together with the process speed N determines the maximum achievable spatial resolution of the correlation signals and also influences the value m_{ij} corresponding to the confidence level of 95% which was selected as threshold for the joint quality evaluation procedure.

The sensor was then calibrated with optimized welding parameters and its response to the occurrence of welds was also verified. Such procedure helped in tuning the sensitivity of the sensor algorithm. The final result we obtained was that the sensor has proven is capability in detecting the most important welding defects such as lack of weld penetration, bead surface oxidation (due to ineffective gas shielding), gaseous inclusions (due to surface pollution for inappropriate choice of the main operating parameters, as shielding gas flux, welding speed, laser power).

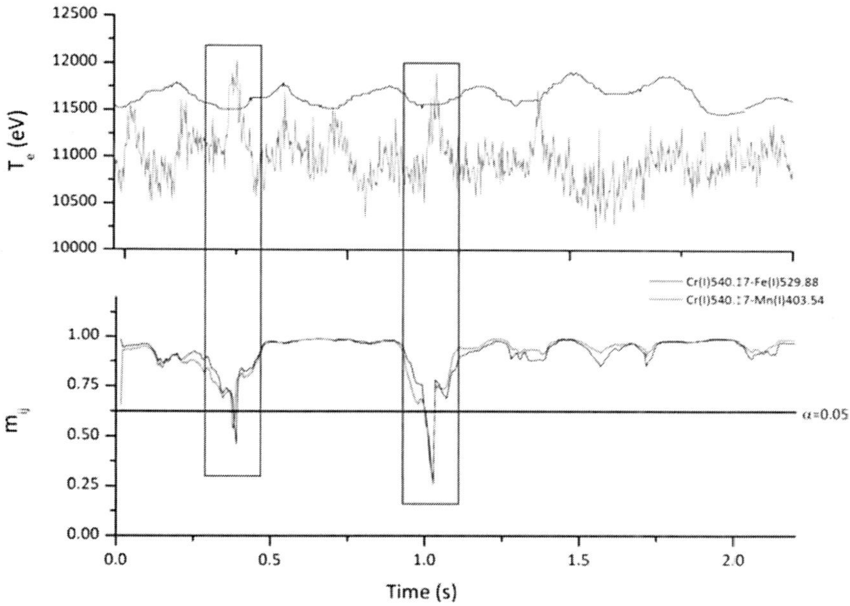

Figure 5. Cr(I) electron temperature and correlation signals for variable welding speed.

Figure 5 illustrates the electron temperature and correlation signals evolution in presence of variable welding speed. It is evident that instabilities in the molten pool as well as in the plasma plume are clearly detected by the sensor. In fact both the electron temperature and the correlation signal values exceed the threshold values.

It is worth mentioning that one of the main advantages of our sensor lies in the relatively easy procedure for the calibration which allows any operator to optimize the values of the main process parameters, like the welding speed or the laser power to obtain sound welded pipes.

4. CONCLUSIONS

In this paper we have briefly reviewed the state of art of optical sensors for laser welding process. Several techniques demonstrate the effectiveness of using optical signals for sensitive, selective and real time defect detection. We also reported on the development of a spectroscopic sensor, patented and now available on the market, based on the acquisition of the optical spectra emitted from the laser generated plasma plume. From the optical spectra the sensor evaluates the electron temperature and analyses the correlation between selected pairs of spectral lines. All the calculations are performed in real-time.

Compared to other optical sensors, the main advantage of this system is that it has a great flexibility upon variation of the welding metal or the joint geometries. In fact once the chemical composition of the alloy is known and the relevant plasma emission lines are identified, only a minor calibration of the software settings is necessary. Provided that the emission line parameters are known, this method is particularly user friendly because it does not require too many calculations and can be easily implemented for real-time temperature monitoring.

ACKNOWLEDGMENTS

This work has been partially carried out with financial support from the European Community (contract FP7-SME-2007-1-222279) under the project CLET – "Closed loop control of the laser welding process through the measurement of plasma".

REFERENCES AND NOTES

1. Ancona, A.; Sibillano, T. Monitoring Laser Welding. In *Real-Time Monitoring of Welding Processes*; Zhang, Y.M., Ed.; Woodhead Publishing Limited: Cambridge England, 2008; pp. 260–287.

2. Tu, J.; Miyamoto, I.; Inoue, T. Characterizing keyhole plasma light emission and plume scattering for monitoring 20kW class CO_2 laser welding processes. *J. Laser Appl.* **2002**, *14*, 146–153.

3. Zhang, X.; Chen, W.; Ashida, E.; Matsuda, F. Relationship between weld quality and optical emissions in underwater Nd:YAG laser welding. *Opt. Laser Eng.* **2004**, *41*, 717–730.

4. Gatzweiler, W.; Maischner, D.; Beyer, E. On-line diagnostics for process control in welding with CO2 lasers. *Proc. SPIE* **1998**,*1020*, 142–148.

5. Park, Y.W.; Park, H.; Rhee, S.; Munjin, K. Real time estimation of CO_2 laser weld quality for automotive industry. *Opt. Laser Tech.* **2002**, *34*, 135–142.

6. Park, H.; Rhee, S. Analysis of mechanism of plasma and spatter in CO_2 laser welding of galvanized steel. *Opt. Laser Tech.* **1999**, *31*, 119–126.

7. Peters, C.N.D.; Jones, J.D.C.; Hand, D. P.; Haran, F.M. Focus control of lasers in material processing operations. *European Pat.* **1997**, *0811453*.

8. Hand, D.P.; Peters, C.; Jones, J.D.C. Nd:YAG laser welding process monitoring by non-intrusive optical detection in the fiber optic delivery system. *Meas. Sci. Technol.* **1995**, *6*, 1389–1394.

9. Haran, F.M.; Hand, D.P.; Jones, J.D.C. Real-time focus control in laser welding. *Meas. Sci. Technol. Rapid Commun.* **1996**, *7*, 1095–1098.

10. Haran, F.M.; Hand, D.P.; Peters, C.; Jones, J.D.C. Focus control system for laser welding. *Appl. Opt.* **1997**, *36*, 5246–5251.

11. Bardin, F.; Cobo, A.; Lopez-Higuera, L.M.; Collin, O.; Aubry, P.; Dubois, T.; Hogstrom, M.; Nylen, P.; Jonsson, P.; Jones, L.D.C.; Hand, D.P. Closed-loop power and focus control of laser welding for full-penetration monitoring. *Appl. Opt.* **2005**, *44*, 13–21.

12. Bardin, F.; Cobo, A.; Lopez-Higuera, L.M.; Collin, O.; Aubry, P.; Dubois, T.; Hogstrom, M.; Nylen, P.; Jonsson, P.; Jones, L.D.C.; Hand, D. P. Optical techniques for real-time penetration monitoring for laser welding. *Appl. Opt.* **2005**, *44*, 3869–3876.

13. Bagger, C.; Olsen, F.O. Laser welding closed.loop power control. *J. Laser Appl.* **2003**, *15*, 19–24.

14. Rockstrok, T.J.; Mazumder, J. Spectroscopic studies of plasma during CW laser material interaction. *J. Appl. Phys.* **1987**, *61*, 917–923.

15. Garcia-Allende, P.B.; Mirapeix, J.; Conde, O.M.; Cobo, A.; López-Higuera, J.M. Spectral processing technique based on feature selection and artificial neural networks for arc-welding quality monitoring. *NDT&E Int.* **2007**, *42*, 56–63.

16. Garcia-Allende, P.B.; Mirapeix, J.; Conde, O.M.; Cobo, A.; Lopez-Higuerak, J.L. Arc-Welding Spectroscopic Monitoring based on Feature Selection and Neural Networks. *Sensors* **2008**, *8*, 6496–6506.

17. Ancona, A.; Spagnolo, V.; Lugarà, P.M.; Ferrara, M. Optical sensor for real-time monitoring of CO_2 laser welding process.*Appl. Opt.* **2001**, *42*, 6019–6025.

18. Sibillano, T.; Ancona, A.; Berardi, V.; Lugarà, P.M. Correlation analysis in laser welding plasma. *Opt. Comm.* **2005**, *251*, 139–148.

19. Sibillano, T.; Ancona, A.; Berardi, V.; Schingaro, E.; Parente, P; Lugarà, P.M. Correlation spectroscopy as a tool for detecting losses of ligand elements in laser welding of aluminium alloys. *Opt. Laser Eng.* **2006**, *44*, 1324–1335.

20. Sibillano, T.; Ancona, A.; Berardi, V.; Lugarà, P.M. Real-time monitoring of laser welding by correlation analysis: the case of AA5083. *Opt. Laser Eng.* **2007**, *45*, 1005–1009.

5

CHAPTER

Crack Propagation Analysis and Rehabilitation by HFMI of Pre-Fatigued Welded Structures

M. Leitner[1], Z. Barsoum[2], F. Schäfers[3]

[1]*Department Product Engineering, Chair of Mechanical Engineering, Montanuniversität Leoben, Leoben, Austria*

[2]*KTH-Royal Institute of Technology, Department of Aeronautical and Vehicle Engineering, Division of Lightweight Structures, Stockholm, Sweden*

[3]*Pitec GmbH, Heudorf, Germany*

ABSTRACT

This paper deals with a crack propagation analysis of welded structures and rehabilitation after pre-fatigue loading by the high-frequency mechanical impact (HFMI) post-treatment technique. The investigated specimen type is a thin-walled longitudinal stiffener made of mild steel *S355*. Fracture mechanical calculations are primarily performed on the basis of the weight-function approach. Thereby, the local residual stress condition at the weld toe is considered by the aid of a structural weld simulation, whereas the numerically evaluated residual stress distribution in depth agrees well to X-ray measurement results. The fracture mechanical analysis illustrates that by incorporating the residual stress state, the calculated lifetime is in good accordance to the conducted fatigue test results. By application of the HFMI-treatment as rehabilitation method, it is found that the beneficial post-treatment effect increases especially by a reduction of the applied load-level. Hence, particularly for minor nominal stress ranges near the high-cycle fatigue region, the mechanical post-treatment as repair method is utmost effective leading to almost equal fatigue strength as for the HFMI-treated specimens without pre-cycling. Finally, proposals for the crack growth assessment of welded structures and a conservative application of HFMI as rehabilitation method for mild steel joints are provided.

Keywords: Crack propagation Fatigue loading Fatigue improvement Weld toes Residual stresses Mechanical stress relief Stress distribution Post weld operations Computation Structural members Fracture mechanics

1 INTRODUCTION

To assess the fatigue strength of welded joints, recommendations [1] are available, which involve different stress-based approaches, such as the nominal, the structural and the effective notch stress method. However, it is well known that the fatigue strength of welded parts is significantly affected by local properties, such as the local geometry at the weld toe or root and the residual stress condition. In addition, discontinuities like flaws, undercuts or pores cannot be avoided during gas-metal arc welding and need to be statistically considered in the fatigue assessment, whereas fracture mechanical procedures based on an evaluation of the linear-elastic crack propagation are well suited for this kind of task [2, 3]. An important application area of this concept occurs in case of large structures, where often additional service life is granted by maintaining recurring service intervals. If a crack is then non-destructively detected [4], different kinds of post-treatment techniques are feasible, at which one efficient method is offered by the high-frequency mechanical impact (HFMI) treatment due to a feasible and expeditious application on site. Recently, numerous investigations [5, 6] showed the beneficial effect of this method for an application without any cyclic pre-loading, but however, the application as rehabilitation purpose is not sufficiently investigated yet.

As a proper assessment of the crack propagation during operation and the specific conditions to extend the lifetime is of great importance, this paper contributes to the crack propagation analysis of welded joints and the subsequent rehabilitation of pre-fatigued structures by the HFMI post-treatment, see chapter 6.2. As the fatigue crack growth regime is predominant in case of notched joints, a longitudinal stiffener specimen made of mild steel *S355* is investigated (Fig. 1).

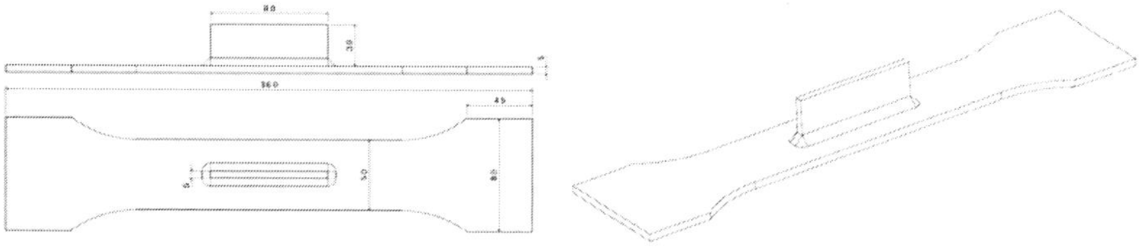

Figure. 1 Geometry of investigated longitudinal stiffener specimen

2 FATIGUE TESTS

At first, fatigue tests in the as-welded and HFMI-treated conditions without any pre-fatigue of the specimens are performed. The stress ratio is set to $R = 0.1$, and the abort criterion is burst fracture. The evaluated nominal a/N curves are presented in Fig. 2exhibiting a survival probability of $P_S = 97.7$ %.

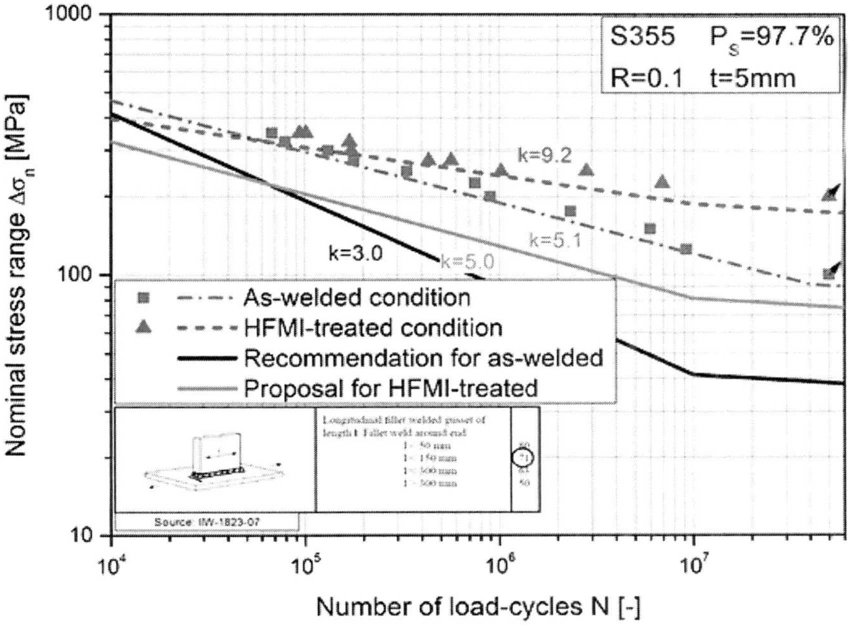

Figure. 2 Nominal a/N curves for as-welded and HFMI-treated conditions

The results show that the HFMI-treatment increases the fatigue strength especially in the high-cycle fatigue region due to a shift of the transition knee point to a lower lifetime. The nominal stress values of the HFMI-treated specimens at the run-out level at fifty million load-cycles exceed with $\Delta\sigma_n = 200$ MPa the as-welded condition by a factor of two. In the finite lifetime regime, an improvement is still observable, mainly due to an increase of the slope from $k = 5.1$ for the as-welded to $k = 9.2$ for the post-treated specimens. A comparison to the recommended [1] curve for the as-welded and the proposed [7] curve for the HFMI-treated condition indicates a high welding and post-treatment quality which fulfil the recommended demands [8].

For an analysis of the technical crack initiation and the subsequent macroscopic crack propagation, strain gauges at the weld toe region of the end-of-seam areas are applied and optical measurements by the aid of a zinc-oxide paste as contrast medium are performed, see Fig. 3. A sensitivity study involving numerous tests leads to the relation between the decrease of the maximum measured strain gauge values at the weld toe, the optically evaluated surface crack, and to the corresponding crack depth based on micrographs of the pre-fatigued specimens. In Fig. 3a, a drop of about 20 % of the initial stress $\sigma_{max,1}$ down to $\sigma_{max,2}$ is observed. The attendant optical evaluation of the surface crack behaviour for the two corresponding fatigue-lives in Fig. 3b shows a crack initiation at $N = 7e4$ and $\sigma_{max,2}$ at which the fatigue testing is afterwards stopped. Detailed information regarding the fatigue tests and strain gauge measurements is provided in [9].

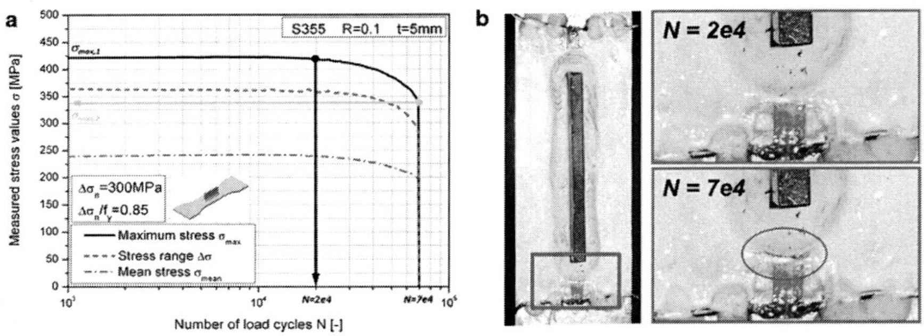

Figure. 3 **a** Strain gauge measurements and **b** optical surface crack detection [9]

Based on the presented experimental procedure, a pre-cycling of the specimens in the as-welded condition up to a defined crack depth is enabled, see also [10]. In Fig. 4, the resulting pre-fatigued a/N curve implying a crack depth of about $a = 1$ mm in comparison to the previously depicted tests up to burst fracture is evaluated by a survival probability of$P_S = 50$ %. Thereby, only a minor difference is noticeable, which is caused by the fact that the bulk of the total lifetime is spent in the small crack size region. Experimental and numerical investigations in [11] show that about 80 % of the total fatigue-life is spent to grow the crack up to 20 % of the related wall thickness, which basically proves the results in this paper. The pre-fatigued specimens are further on used to assess the potential of the HFMI treatment as rehabilitation method.

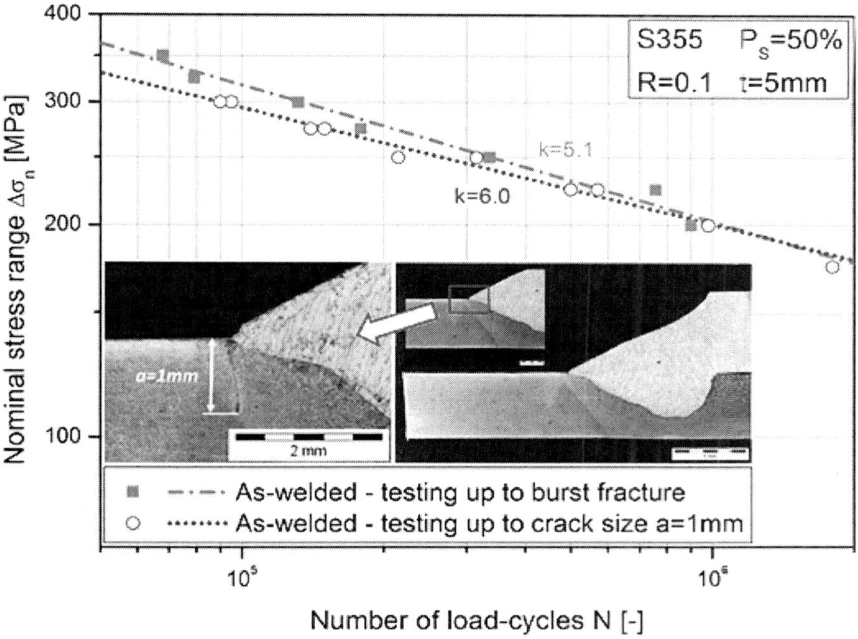

Figure. 4 Comparison of test results with specimens up to burst fracture and crack size $a = 1$ mm

3 LOCAL STRESS CONDITION

For the evaluation of the local notch stress distribution in depth at the weld toe, a quarter-symmetric model with a hexahedral element mesh and a unified tension load is set-up. The modelled geometry depicts the nominal cross section of the specimens, and local deviations from the design geometry are not incorporated in detail. Distortion measurements of the specimens before fatigue testing exhibit only minor values up to one-tenth-degree angular deformation. A further evaluation of the influence factor k_m given in [1] shows almost no impact, and therefore, an explicit consideration in the numerical analysis is not necessary. The mesh is built-up in accordance to the recommended values for the effective notch stress assessment in [12] and is adapted to study the effect of varying weld toe radii. The stress evaluation is performed for the component *S11* appearing in loading direction of the specimen and therefore most significant for the subsequent crack propagation analysis, see Fig. 5a.

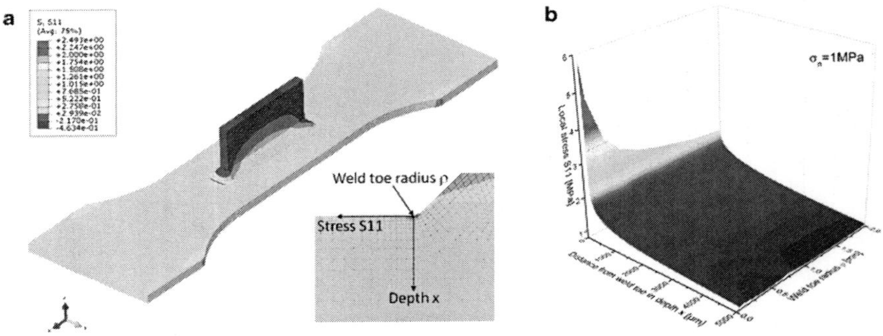

Figure. 5 a Numerical model for analysis of local stress condition and **b** in-depth notch stress distribution in dependency of weld toe radius

In Fig. 5b, the results of the numerical stress analysis are depicted. Thereby, the local stress distribution in depth in dependency of the weld toe radius, ranging from ρ = 0.05 to 2 mm, is presented. It is shown that at the surface region up to a depth of about 0.1 mm, the weld toe radius has a major influence on the local stress condition. Especially for small radii of about ρ = 0.05 up to 0.2 mm, a high stress concentration is evaluated.

4 STRUCTURAL WELD SIMULATION

In order to improve the accuracy of the crack propagation results, also the residual stress condition needs to be incorporated. In general, residual stresses can be determined by means of measurements or numerical analysis. Basic investigations involving a structural weld simulation of the investigated joint type and material are shown in [13]. Thereby, an enhanced method to incorporate the transient work-hardening effect of different material phases into a thermo-mechanically coupled structural weld process simulation is presented, whereas the work-hardening behaviour of metals principally differs for soft ferritic-perlitic and hard bainitic, or even martensitic, phases. For the conducted thermo-mechanical coupled weld process simulation, the software package Sysweld [14] is utilized, which supports the implementation of different work-hardening models, such as the basic isotropic behaviour and the kinematic hardening rule according to [15]. Moreover, it is possible to apply the kinematic portion [16], which takes, besides the effect of backstress, the influence of dynamic recovery during cyclic loading into account [17].

One of the main drawbacks of the described material models is that they are basically not able to consider the effect of phase-change during the transient weld simulation. This means, that the unique material property database is defined at the stage of input card definition and cannot be easily modified for each element possessing process time- and temperature-dependent different hardening parameters, see [13].

To solve the task of transient change in work-hardening per element during the numerical simulation process, a user-defined SIL routine was previously developed [18]. It enables the implementation of a temperature- and phase-dependent work-hardening behaviour during the transient thermo-mechanical weld simulation. Based on low-cycle-fatigue test results, three different work-hardening behaviours are classified, which depend on the temperature and the major phase content, separated in mainly ferritic-perlitic as soft microstructure; predominantly bainitic-martensitic as hard phase or mixed microstructure. At ambient temperature, the kinematic behaviour is more appropriate, whereas at higher temperatures, close to the austenitic region, the isotropic work-hardening rule is more applicable. The results of the mixed material model calculate the residual stress condition quite well [13]. These results are applied for the investigations in this paper. In Fig. 6a, the conducted structural weld simulation model and the temperature distribution during welding is displayed.

In Fig. 6b, the numerically evaluated residual stress values *S11* in depth are compared to experimental data by X-ray measurements for the as-welded condition. Additional residual stress measurements in [9] focusing on the base material exhibit a major compressive residual stress state in the surface layer of the metal sheet, which primarily emerge due to the base-material rolling process. By a superposition of these compressive stresses in the numerical weld simulation, the results in the as-welded condition show a good compliance to the measured values and are therefore used for the subsequent crack propagation analysis.

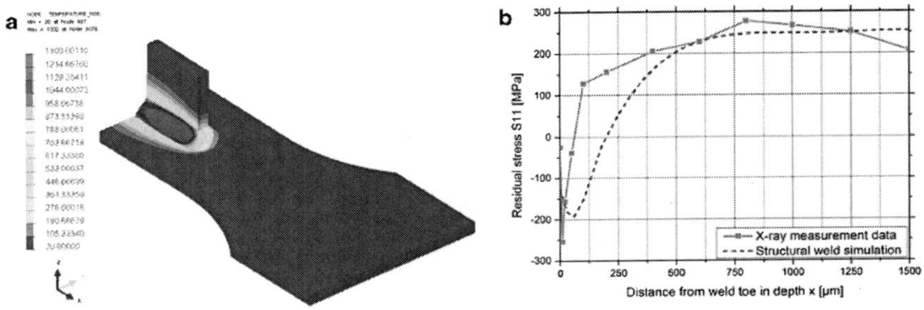

Figure. 6 a Temperature distribution during welding and **b** comparison of measured and simulated residual stress values

5 CRACK PROPAGATION ANALYSIS

Fracture mechanical approaches provide the basis for the assessment of cracks or crack-like imperfections, where the crack propagation significantly exceeds the crack initiation phase, see [19]. For welded joints, the basic elements of the linear-elastic crack growth calculation are the stress intensity factor *K*, the crack shape and path and the chosen material parameters, see Table 1. The crack propagation is mostly calculated based on the simple power law according to Paris and Erdogan [21]:

$$\frac{da}{dN} = C \cdot \Delta K^m \text{ for } \Delta K > \Delta K_{th} \tag{1}$$

Table 1 Recommended crack propagation data for steel [20]

Paris power law parameters (charact. values)	Threshold values ΔK_{th} [MPa$\sqrt{m}$]			
	$R > 0.5$	$0 \leq R \leq 0.5$	$R < 0$	Surface crack depth < 1 mm
$C = 9.5 \cdot 10^{-12}$ $m = 3.0$	2.0	$5.4 - 6.8 \cdot R$	5.4	≤ 2.0

For standardized crack geometries and load configurations, numerous stress intensity factor formulae are available [22]. However, there are basically no ready-made solutions for welded structures except for a few crack configurations. Therefore, the weight function technique [23, 24] was employed in order to determine stress intensity factors for cracks in real welded structures, see [25].

5.1 Weight-function approach

The weight function approach is basically defined as the stress intensity factor *K* induced by the simplest load configuration, such as a pair of unit forces *F* attached to the crack surface, see Fig. 7. The operation of each differential force on a crack is specified by the weight-function [20, 25].

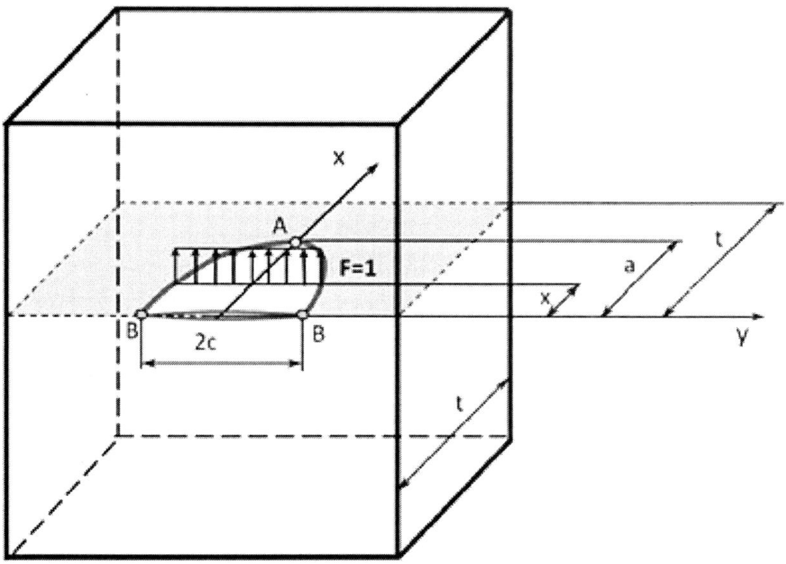

Figure. 7 Semi-elliptical surface crack in plate [26]

A general form to describe the weight function is provided by [24] according to Eq. 2. The containing parameters M_1, M_2, and M_3 depend on the actual crack geometry and are listed in [24, 25] for numerous crack configurations. For the crack propagation analysis of the investigated longitudinal stiffener in this paper, the accordant expressions for a semi-elliptical crack at the deepest point A are applied.

$$m(x,a) = \frac{2\cdot F}{\sqrt{2\cdot\pi\cdot(a-x)}}\cdot\left[1 + M_1\cdot\left(1-\frac{x}{a}\right)^{\frac{1}{2}} + M_2\cdot\left(1-\frac{x}{a}\right) + M_3\cdot\left(1-\frac{x}{a}\right)^{\frac{3}{2}}\right] \tag{2}$$

To calculate the stress intensity factor *K*, the knowledge of the stress distribution $\sigma(x)$ in the prospective crack plane of the un-cracked plate is required. In [20], it is recommended to involve the stress distribution assuming a weld toe radius of $\rho = 0$ to 0.2 mm, which is already computed within the local stress analysis depicted in "Local stress condition". Further on, the stress intensity factor *K* is calculated by integrating the product of the stress distribution $\sigma(x)$ and the weight function $m(x,a)$ for each crack increment during cyclic loading, see Eq. 3.

$$K = \int_0^a \sigma(x)\cdot m(x,a)\cdot dx \tag{3}$$

To calculate the fatigue-life more precisely, the influence of the residual stress condition at the vicinity of the crack at the weld toe has to be considered. Based on the numerically evaluated residual stress condition, a procedure based on investigations in [27] involving an effective stress ratio R_{eff} [28] at the crack tip, is applied, see Eq. 4.

$$\frac{da}{dN} = C\cdot\left(\frac{\Delta K}{1.5-R_{eff}}\right)^m \tag{4}$$

The stress intensity factor range ΔK is defined by the difference of the maximum and minimum stress intensity factors at the crack tip and is hereby not affected by the residual stress state, see Eq. 5.

$$\Delta K = K_{max}-K_{min} \text{ with } K_{max/min} = \int_0^a \sigma_{max/min}(x)\cdot m(x,a)\cdot dx \tag{5}$$

In accordance to the loading-dependent stress intensity factors K_{max} and K_{min}, the residual stress intensity factor K_{res} is calculated in the same way, see Eq. 6. The residual stress distribution in depth $\sigma_{res}(x)$ is derived from the result of the structural weld simulation.

$$K_{res} = \int_0^a \sigma_{res}(x) \cdot m(x, a) \cdot dx \qquad (6)$$

Finally, the effective stress ratio R_{eff} is defined as shown in Eq. 7. Thereby, the local stress ratio at the crack tip is influenced by the local residual stress condition and in this way incorporated in the crack propagation analysis and fatigue-life assessment.

$$R_{eff} = \frac{K_{min} + K_{res}}{K_{max} + K_{res}} \qquad (7)$$

Fracture mechanical calculations are performed in accordance to the presented weight-function procedure for the longitudinal stiffener specimen. For the crack shape, the aspect ratio a/c is one important influence factor, whereas general applicable recommendations are provided in [20]. As a conservative approach, a ratio of $a/c = 0.1$ may be defined, but the calculation results exhibited that the computed lifetime is significantly lower than the fatigue test results. Therefore, the proposed formula for ends of longitudinal stiffeners in [20] is applied (Eq. 8), leading to a/c values of 0.7 to 0.8, which are also in good accordance to the test results of the fractured specimens.

$$c = \frac{6.71 + 2.58 \cdot a}{2} \qquad (8)$$

A comparison of the calculated N_{calc} and experimentally evaluated load-cycles N_{exp} is presented in Fig. 8 as normalized ratio N_{calc} / N_{exp} in dependency of the initial crack length a_i and the weld toe radius ρ. Basically, it is shown that not only the initial crack length value of $a_i = 0.1$ mm is generally well applicable but also values of $a_i = 0.05$ mm which are suggested for mechanical engineering applications or $a_i = 0.15$ mm as a conservative approach by fitness for purpose codes [20] are acceptable. However, to analyse the influence of the initial crack length in detail, a variation of $a_i = 0.05$ up to $a_i = 0.2$ mm is included. In the finite lifetime regime at a nominal stress ratio of $\Delta\sigma_n = 200$ MPa, the calculated load-cycles are affected by the weld toe radius only in case of minor initial crack lengths of $a_i = 0.05$ mm, leading to a significant overestimation of the lifetime, see Fig. 8a. For higher initial crack values, the influence of the weld toe radius is consequently decreasing.

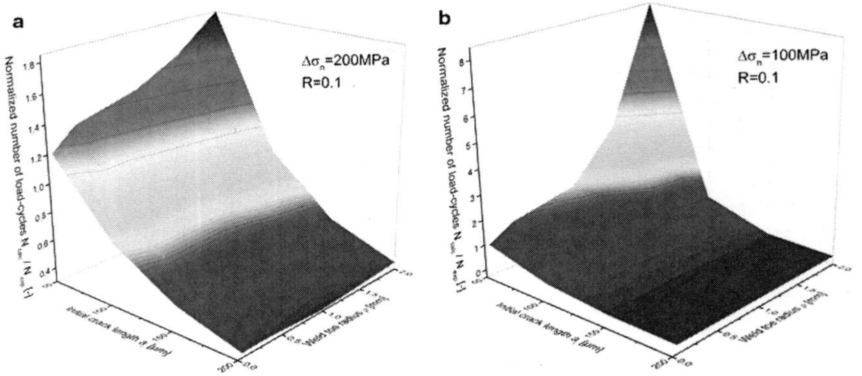

Figure. 8 Normalized number of load-cycles N_{calc} / N_{exp} at a nominal stress range of **a** $\Delta\sigma_n = 200$ MPa and **b** $\Delta\sigma_n = 100$ MPa

On the contrary, in the high-cycle fatigue region at a nominal stress ratio of $\Delta\sigma_n = 100$ MPa, the minimum investigated initial crack length tends to result in good accordance to the fatigue tests incorporating weld toe radii smaller than $\rho = 0.2$ mm, see Fig. 8b. Assuming a higher notch radius, the calculated lifetime is overestimated up to a factor of eight and is, therefore, not applicable. Summarized, the calculated lifetime N_{calc} is strongly influenced by the weld toe radius, especially at a minor defined initial crack length of about $a_i = 0.05$ mm.

As the calculated lifetime is significantly affected by the initial crack length and the weld toe radius at the high-cycle fatigue region, a detailed analysis of the crack tip loading involving a nominal stress of $\Delta\sigma_n = 100$ MPa is performed, see Fig. 9. Thereby, the influence of the weld toe radius on the stress intensity factor range ΔK in dependency of the actual crack length a is quite small, see Fig. 9a. Especially if the crack is technically propagated and not influenced by the stress concentration at the weld toe region anymore, the impact of the weld toe radius and the associated stress distribution on the stress intensity factor is negligible. Figure 9b displays the stress intensity factor range ΔK related to the minimum recommended [20] threshold value of $\Delta K_{th,min} = 2$ MPa$\sqrt{m}$ in dependence of the initial crack length a_i and the weld toe radius ρ. One can see that especially for minor values of $a_i < 0.1$ mm, the acting ΔK is below the threshold value and therefore, basically no long-crack-based crack propagation occurs.

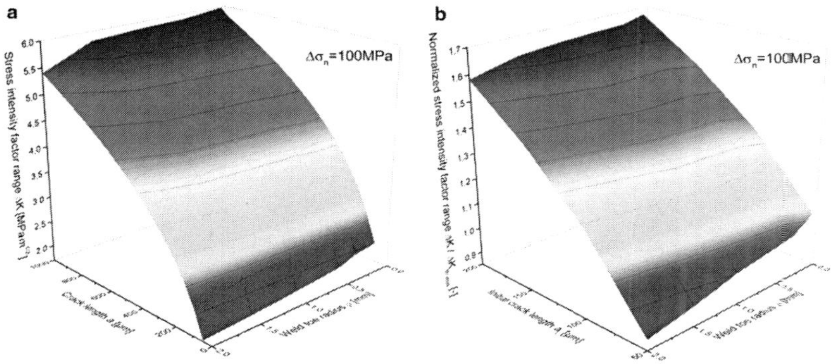

Figure. 9 **a** Stress intensity factor range ΔK during crack propagation and **b** ratio of ΔK / $\Delta K_{th,min}$ in dependency of initial crack length a_i

Summarized, an initial crack length of a_i = 0.1 mm and a weld toe radius of ρ = 0.05 mm lead to good accordance of the fatigue test results and are also in agreement with the recommended values [20]. These values are further on applied for the successive crack propagation analysis.

Figure 10a shows the effective stress ratio R_{eff} at the crack tip in dependence of the stress ratio R due to the loading and the actual crack length a. As the influence of the residual stress condition is considered by R_{eff}, the effect of the residual stress distribution is clearly observable for the lowest stress ratio of R = −1. In this case, the compressive residual stress state at the surface layer of the weld toe region affects the local stress ratio R_{eff} in a major way. If the crack propagates, the effect is reduced and at a crack depth of about a = 0.5 mm, no more influence is observable.

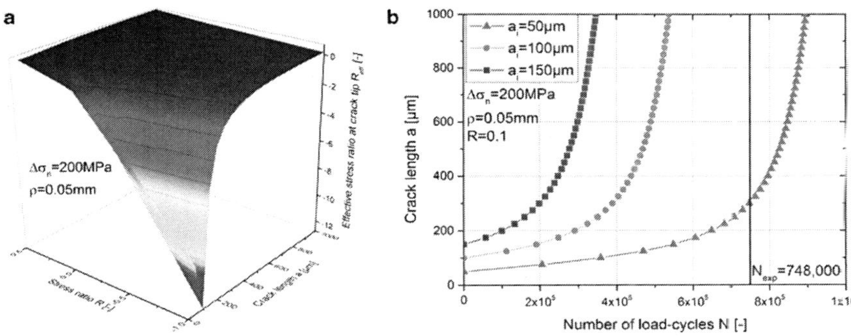

Figure. 10 **a**. Effective stress ratio R_{eff} at crack tip and **b** calculated a/N curves for different initial crack lengths a_i

Otherwise, for high applied tumescent load-ratios of $R = 0.5$, there is almost no effect of the residual stress condition on the local stress ratio at the crack tip. This is in accordance with the IIW-recommendation, whereas a fatigue testing at high mean stresses, preferably at $R = 0.5$, should be conducted in order to minimize the influence of the weld process-induced residual stress condition.

In Fig. 10b, the resulting a/N curves for a nominal stress range of $\Delta\sigma_n = 200$ MPa and the applied stress ratio of $R = 0.1$ are depicted for different initial crack lengths a_i. As already stated, an assumption of $a_i = 0.05$ mm may lead to an overestimation of the lifetime, which is also evaluated in this case. The recommended value of $a_i = 0.1$ mm calculates the lifetime in a safe but appropriate way, and further on, a definition of $a_i = 0.15$ mm leads, as expected, to a conservative fatigue-life assessment.

5.2 Parametric equation approach

In case of simple weld geometries, such as butt welds, T-joints or longitudinal stiffeners, parametric formulae are provided [20, 29]. Thereby, the stress intensity factor K is calculated according to Eq. 9 considering the crack-geometry-dependent correction function $Y(a)$ and an additional adjustment for the local notch of the weld toe $M_k(a)$. In [30], the magnification factor M_k is presented, leading to a comparable stress condition for a crack of the same geometry, but without the presence of the weld notch. Furthermore, a separation of membrane σ_m and shell bending stress σ_b is necessary for most of the parametric formulae, whereas the resulting stress intensity factor K is computed as a superposition of both stress components.

$$K = \sqrt{\pi \cdot a} \cdot \left[\sigma_m \cdot Y_m(a) \cdot M_{k,m}(a) + \sigma_b \cdot Y_b(a) \cdot M_{k,b}(a)\right] \quad (9)$$

In [20], the stress magnification factor M_k for the end-of-seam region of a longitudinal stiffener is defined based on Eq. 10. Herein, the factor C and exponent k depend on the dimensions of the corresponding weld detail.

$$M_k = C \cdot \left(\frac{a}{t}\right)^k \quad (10)$$

Recent studies [31] involving a comparison of analytical, numerical and experimental results show good accordance and the basic applicability of the presented parametric formulae.

5.3 Comparison of results

To validate the fracture mechanical calculation, the calculated lifetime values are compared to the experimental fatigue test results, see Table 2. Thereby, the initial crack length is set to the recommended value of $a_i = 0.1$ mm and the calculations are performed up to a final crack length of $a_f = 1$ mm because as shown in Fig. 4, within this crack propagation region, most of the lifetime is spent and just a minor remaining lifetime up to burst fracture is evaluated.

Table 2 Evaluated lifetime *N* for different load-levels

Initial crack length $a_i = 0.1$ mm				
Nominal stress range $\Delta\sigma_n$ [MPa]	N_{exp} [−]	N_{WF} [−] (incl. RS)	N_{WF} [−] (excl. RS)	N_{PE} [−] (excl. RS)
100	2.1e7	1.5e7	1.3e6	194,000
200	748,000	537,000	164,000	24,000
300	95,000	100,000	48,000	7000

Firstly, a comparison of the experimental lifetime N_{exp} and the calculated load-cycles by the weight-function approach N_{WF} shows that by including the residual stress (RS) condition a well-matching fatigue assessment is deduced. On the contrary, the exclusion of the residual stress state leads to a significant reduction of the calculated lifetime, which can be explained by the mainly compressive residual stresses at the surface region maintaining a reduced crack propagation rate and, therefore, an increased lifetime. The effect of a super-positioned residual stress field on the crack propagation of surface cracks is principally analysed in [32].

Secondly, the application of the parametric equation approach to assess the fatigue-life exhibits the lowest and most conservative lifetime values N_{PE}. Also in this case, the beneficial compressive residual stresses are not considered, conducting to a major decrease for the lifetime estimation. However, one advantage of this method is the simple applicability with less effort and expenditure of time.

6 REHABILITATION OF PRE-FATIGUED STRUCTURES

6.1 Methods

If a crack is detected or the service time of the structure or component is depleted, several rehabilitation methods are applicable. Simple retrofitting techniques are to attach bolted splices to the cracked parts or to achieve crack-growth arrest by drilling sop holes at the end of cracks [33]. A more elaborated way to restore cracked areas is fulfilled by re-welding. Especially wear-damaged sections of supporting structures are often weld-repaired, whereas an influence by the tri- or bi-metal system on the fatigue crack growth should be analysed [34]. Another weld-repair technique which also improves the fatigue performance is based on the application of low-temperature transformation welding wires. Thereby, compressive residual stress fields in the vicinity of the weld bead are introduced which has a beneficial effect on crack initiation and propagation and therefore enhances the fatigue strength [35].

Further technologies to rehabilitate pre-fatigued structures include the application of post-treatment methods. In [36], tungsten inert gas (TIG) and plasma dressing are applied to repair crack-containing T-welded joints. It is shown that by TIG-dressing, only crack depths up to 2.5 mm can be successfully repaired. By the aid of plasma dressing, the results show an effect also for greater crack depths up to 4 mm, whereas the fatigue-lives were found to be similar as for the as-welded condition. An overview of further available repair methods like the local explosive treatment, local heat treatment and high-frequency mechanical impact (HFMI) treatment with traditional techniques is given in [37]. However, based on the applicability and industrial feasibility, the HFMI treatment is one utmost efficient method, which is further on exemplified.

6.2 HFMI treatment

Not only for un-cracked but also for pre-fatigued joints, recent investigations [10, 38, 39] show a great beneficial effect for rehabilitation by the HFMI treatment, whereas the fatigue strength of the repaired specimens is equal or even exceeds the HFMI-treated condition without pre-cycling. As described, pre-fatigued specimens with a crack depth of $a \sim 1$ mm are analysed to assess the potential of HFMI as a rehabilitation method without additional re-welding. A comparison of the evaluated fatigue test results is shown in Fig. 11. Thereby, the fatigue enhancement by the rehabilitation in the finite lifetime regime basically increases with a decreasing nominal load stress range. In the high-cycle fatigue region, the fatigue strength at the run-out level is almost identical to the HFMI-treated curve without pre-fatigue. Detailed information about the HFMI-treatment and fatigue tests is provided in [9].

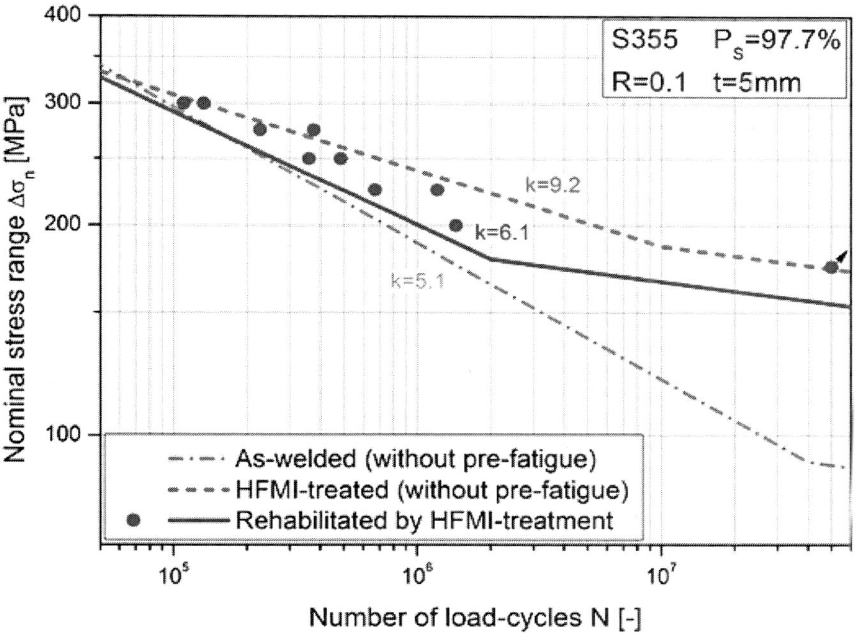

Figure. 11 a/N curve of HFMI-rehabilitated specimens

The results are in good agreement with the previous investigations in literature, which generally confirm the applicability of the HFMI treatment as rehabilitation technique. One main influencing factor on the effectiveness is the crack length before performing the HFMI rehabilitation. In [10], it is shown, that for crack lengths above 1 to 2 mm, a significant reduction of the benefit occurs, which is reasoned by the limited penetration depth by the HFMI treatment, depending on the local notch geometry, material properties and post-treatment parameters.

For the quantification of the rehabilitation benefit by the mechanical post-treatment, an advantageous gain factor g_N is introduced in [10]. The value is defined as the ratio of the fatigue-life of the repaired joint N_r to the expected fatigue-life of the joint if it was not repaired N_e, see Eqs. 11 and 12.

$$N_r = N_i + N_{bh} + N_{aH} \tag{11}$$

$$N_e = N_i + N_{bh} + N'_{aH} \tag{12}$$

Thereby, N_i is the crack initiation life before the first crack for repair starts, N_{bh} is the fatigue-life in crack propagation, N_{aH} is the fatigue-life after repair including crack initiation and propagation stage, and N'_{aH} is the fatigue-life that would be obtained if the HFMI-repair-treatment would not be performed. An evaluation of the gain factor g_N based on the experiments and accompanying crack initiation and propagation measurements indicates that the beneficial effect for a constant pre-crack length is depending on the applied load level, whereas for a higher nominal stress at $\Delta\sigma_n$ = 300 MPa, the assessment leads to g_N = 1.01 and for a lower nominal stress at $\Delta\sigma_n$ = 200 MPa to g_N = 1.60. The evaluated values are basically in good accordance to the results presented in [10]. Figure 12 presents an overview of the fatigue test data for HFMI-rehabilitated joints made of mild steel S355 considering the results of [10, 39] and this work (see Appendix). The comparison clearly shows that the beneficial effect of the HFMI treatment for pre-fatigued joints increases with a decrease of the nominal load.

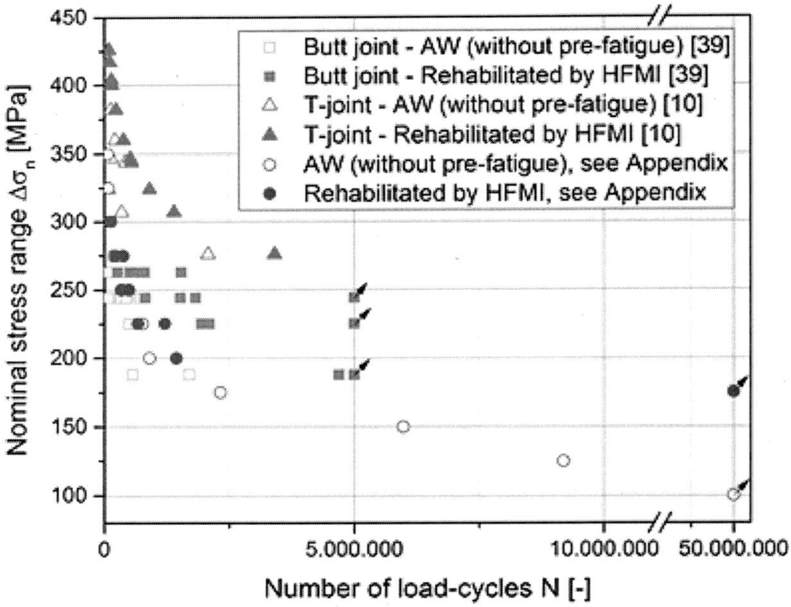

Figure. 12 Overview of fatigue test data for HFMI-rehabilitated mild steel joints ($R = 0.1$)

Based on the investigated butt joint series in [39], it is concluded that a lifetime extension by the HFMI treatment is expected up to a maximum crack depth of $a = 0.5$ mm. In [10], different initial crack depths and their effect on the total lifetime after rehabilitation by HFMI is investigated, showing that the benefit is greater in repaired fatigue cracks with depths of $a \leq 2.5$ mm. However, several influences, such as structural weld detail, loading type, mean stress state, HFMI treatment parameters and compressive residual stress condition in depth, influence the effectiveness of the lifetime extension by HFMI. Considering the results in [10, 39] and by this work, a conservative maximum crack depth of $a_{max} = 0.5$ mm can be recommended leading to an extension by the HFMI rehabilitation. In case of complex welded structures, an exact measurement of the crack length is not always possible maintaining an uncertainty of the applicability of the HFMI treatment. As exploited in [40], a re-welding and additional HFMI post-treatment as fatigue strengthening rehabilitation procedure is recommendable.

7 CONCLUSION

Crack propagation analysis of the investigated longitudinal stiffener based on the application of the weight-function approach and considering the local residual stress condition leads to a good agreement with the evaluated fatigue test results. On the contrary, by neglecting the residual stress state, the deviation significantly increases towards a conservative design. The effect of weld process-induced residual stresses is essential for the analysed load-applied stress ratio of $R = 0.1$ but diminishes at higher mean stress conditions. Additional numerical studies show a great influence of the weld toe radius, whereas an applied value of $\rho = 0.05$ mm exhibits appropriate results of the fracture mechanical calculations and matches well with the recommendation. A final comparison to the fracture mechanical standard approach based on parametric formulae shows that this method leads to a conservative assessment of the fatigue-life.

Fatigue test results in as-welded, HFMI-treated and pre-cycled condition up to a crack depth of $a \sim 1$ mm and subsequent HFMI repair indicate a beneficial effect of the mechanical post-treatment for pre-cracked joints. The fatigue strength after rehabilitation generally depends on the crack length before repair and furthermore on the applied load level, whereas the benefit increases with a decrease of the nominal stress range. Based on the own investigations and results within this work and by consideration of previously analyses regarding fatigue crack growth assessment and rehabilitation of pre-fatigued structures, the following conclusions can be drawn:

- The weight-function approach including an incorporation of the local residual stress condition is well applicable to assess the crack growth of welded joints based on the recommended values with a weld toe radius of $\rho = 0.05$ mm and an initial crack size of $a_i = 0.1$ mm.
- An estimation of the crack size based on parametric formulae to calculate the stress intensity factor range ΔK shows a conservative fatigue assessment leading to an underrating of the lifetime up to a certain final crack length.
- On the basis of the results within this work, it can be concluded that if a structure is pre-fatigued and a certain crack size is detected, the remaining service life can be accurately and efficiently assessed based on the fracture mechanical weight-function approach.
- For a rehabilitation of pre-fatigued structures, the applied method significantly depends on the actual crack size. In general, a rehabilitation procedure involving grinding, re-welding and a final additional post-weld treatment is recommended. However, due to a limited time and cost frame, this elaborated routine cannot be completely fulfilled in industrial applications in some cases.
- If the HFMI treatment is applied without previous grinding and re-welding of the weld toe, a maximum crack size in depth of $a_{max} = 0.5$ mm acts as conservative proposal for mild steel joints. For an application up to this crack depth value, the rehabilitation by HFMI leads to an improvement in fatigue strength both in the finite-life and high-cycle fatigue region. Due to slightly varying user-dependent treatment parameters and difference in obtained work-hardening, the applicability of this suggested maximum crack size cannot be thoroughly given for high-strength steels without any further specific investigations by the customer.

ACKNOWLEDGMENTS

Financial support by the Austrian Federal Government (in particular from Bundesministerium für Verkehr, Innovation und Technologie and Bundesministerium für Wirtschaft, Familie und Jugend) represented by Österreichische Forschungsförderungsgesellschaft mbH and the Styrian and the Tyrolean Provincial Government, represented by SteirischeWirtschaftsförderungsgesellschaft mbH and Standortagentur Tirol, within the framework of the COMET Funding Programme is gratefully acknowledged. Open access funding provided by Montanuniversität Leoben.

APPENDIX

Table 3 Fatigue test data points of investigated a/N-curves

As-welded (without pre-fatigue)		**HFMI-treated (without pre-fatigue)**		**Rehabilitated by HFMI treatment**	
$\Delta\sigma_n$ [MPa]	**N [−]**	**$\Delta\sigma_n$ [MPa]**	**N [−]**	**$\Delta\sigma_n$ [MPa]**	**N [−]**
350	67,808	350	101,314	300	133,000
325	79,194	350	93,785	300	110,000
300	131,791	325	169,677	275	377,000
275	178,647	300	172,723	275	226,000
250	335,957	275	565,548	250	485,000
225	755,950	275	434,341	250	360,000
200	900,086	250	2.81E6	225	1.21E6
175	2.32E6	250	1.02E6	225	670,000

As-welded (without pre-fatigue)		**HFMI-treated (without pre-fatigue)**		**Rehabilitated by HFMI treatment**	
$\Delta\sigma_n$ **[MPa]**	N **[−]**	$\Delta\sigma_n$ **[MPa]**	N **[−]**	$\Delta\sigma_n$ **[MPa]**	N **[−]**
150	5.98E6	225	6.90E6	200	1.44E6
125	9.18E6	200	5E7	175	5E7
100	5E7				

REFERENCES

1 Hobbacher A (2009) IIW recommendations for fatigue design of welded joints and components, WRC bulletin 520. The Welding Research Council, New York

2 Hobbacher A (2011) Fracture and fatigue of welded joints and structures. Woodhead Publishing, Cambridge, pp. 91–112.http://www.sciencedirect.com/science/book/9781845695132

3 Zerbst U, Ainsworth RA, Beier HT, Pisarski H, Zhang ZL, Nikbin K, Nitschke-Pagel T, Münstermann S, Kucharczyk P, Klingbeil D (2014) Review on fracture and crack propagation in weldments—a fracture mechanics perspective. Eng Fract Mech 132:200–276

4 Ditchburn RJ, Burke SK, Scala CM (1996) NDT of welds: state of the art. NDT&E Int 29:111–117

5 Yildirim HC, Marquis GB (2012) Overview of fatigue data for high frequency mechanical impact treated welded joints. Weld World 56(7/8):82–96

6 Leitner M, Stoschka M, Eichlseder W (2014) Fatigue enhancement of thin-walled, high-strength steel joints by high-frequency mechanical impact treatment. Weld World 58:29–39

7 Marquis GB, Mikkola E, Yildirim HC, Barsoum Z (2013) Fatigue strength improvement of steel structures by high-frequency mechanical impact: proposed fatigue assessment guidelines. Weld World 57:803–822

8 Marquis GB, Barsoum Z (2014) Fatigue strength improvement of steel structures by high-frequency mechanical impact: proposed procedures and quality assurance guidelines. Weld World 58:19–28

9 Leitner M (2013) Local fatigue assessment of welded and high frequency mechanical impact treated joints, PhD-thesis, Montanuniversität Leoben

10 Branco CM, Infante V, Baptista R (2004) Fatigue behaviour of welded joints with cracks, repaired by hammer peening. Fatigue Fract Eng Mater Struct 27:785–798

11 Shah F (2015) Crack propagation analysis of welded joints by numerical and experimental investigations, Master thesis, Montanuniversität Leoben

12 Fricke W (2008) Guideline for the fatigue assessment by notch stress analysis for welded structures, IIW-Document XIII-2240r1-08/XV-1289r1-08

13 Stoschka M, Ottersböck MJ, Leitner M (2014) Integration of phase-dependent work-hardening into transient weld simulation, Proceedings of the Ninth International Conference on Engineering Computational Technology, Paper 39

14 ESI Group. (2014) SYSWELD Toolbox, release 16

15 Prager W (1949) Recent development in the mathematical theory of plasticity J Appl Phys vol. 235, issue 7

16 Armstrong P, Frederick C (1966) A mathematical representation of the multiaxial Bauschinger effect, CEGB Report RD/B/N731

17 Chaboche J-L (2008) A review of some plasticity and viscoplasticity constitutive theories. Int J Plast 24:1642–1693

18 Ottersböck M, Stoschka M, Thaler M (2013) Study of kinematic strain hardening law in transient welding simulation. Math Model Weld Phenom 10:255–266

19 Fricke W (2003) Fatigue analysis of welded joints: state of development. Mar Struct 16:185–200

20 Hobbacher A (2012) Update of the fracture mechanics chapters of the IIW fatigue design recommendations, IIW-document XIII-2370r1-11 / XV-1376r1-11

21 Paris P, Erdogan F (1963) A critical analysis of crack propagation laws, Journal of Basic Engineering. Trans Am Soc Mech Eng 85(4):528–534

22 Murakami Y.: Stress intensity factors handbook, Pergamon, Oxford, vol. 2, 1987

23 Bueckner HF (1970) A novel principle for the computation of stress intensity factors. Z Angew Math Mech 50:529–546

24 Glinka G, Shen G (1991) Universal features of weight functions for cracks in mode I. Eng Fract Mech 40(6):1135–1146

25 Chattopadhyay A, Glinka G, El-Zein M, Qian J, Formas R (2011) Stress analysis and fatigue of welded structures. Weld World 55(7/8):2–21

26 Goyal R, Glinka G (2013) Fracture mechanics-based estimation of fatigue lives of welded joints. Weld World 57:625–634

27 Glinka G.: Effect of residual stresses on fatigue crack growth in steel weldments under constant and variable amplitude loading, Fracture Mechanics, ASTM STP 677, American Society for Testing and Materials, pp. 198–214, 1979

28 Kurihara M, Katoh A, Kwaahara M (1986) Analysis on fatigue crack growth rates under a wide range of stress ratio. J Press Vessel Technol, Trans ASME 108(2):209–213

29 Hobbacher A (1993) Stress intensity factors of welded joints. Engineering Fracture Mechanics 46(2):173–182, and vol. 49, no. 2, p. 323, 1994

30 Maddox SJ (1975) An analysis of fatigue cracks in fillet welded joints. Int J Fract 11(2):221–243

31 Al-Mukhtar AM, Henkel S, Biermann H, Hübner P (2009) A finite element calculation of stress intensity factors of cruciform and butt welded joints for some geometrical parameters. Jordan J Mech Ind Eng 3(4):236–245

32 Mochizuki M, Miyazaki K (2006) Surface crack propagation analysis under residual stress field. Weld World 50(5/6):38–45

33 Miki C (2008) Retrofitting engineering for fatigue damaged steel structures, IIW-document XIII-2284-09 / WG5-74-07, revised 2008

34 Zhang C, van der Vyver S, Hua X, Lu P (2011) Fatigue crack growth behavior in weld-repaired high-strength low-alloy steel. Eng Fract Mech 78:1862–1875

35 Miki C, Hanji T, Tokunaga K (2012) Weld repair for fatigue-cracked joints in steel bridges by applying low temperature transformation welding wire. Weld World 56(3/4):40–50

36 Ramalho A, Ferreira J, Branco C (2011) Fatigue behaviour of T welded joints rehabilitated by tungsten inert gas and plasma dressing. Mater Des 32:4705–4713

37 Kudryavtsev Y., Kleiman J., Knysh V., Mikheev P.: Fatigue life improvement of structural elements with fatigue cracks, Proceedings of the SEM Annual Conference & Exposition on Experimental and Applied Mechanics, 2002

38 Kudryavtsev Y, Kleiman J, Lobanov L, Knysh V, Voitenko O, Prokopenko G, Lugovskoy A (2007) Rehabilitation and repair of welded elements and structures by ultrasonic peening. Weld World 51(7/8):47–43

39 Weich I (2009) Ermüdungsverhalten mechanisch nachbehandelter Schweißverbindungen in Abhängigkeit des Randschichtzustands (Fatigue behaviour of mechanical post weld treated welds depending on the edge layer condition), PhD-thesis, Technische Universität Braunschweig, in German

40 Tominaga T, Matsuoka K, Sato Y, Suzuki T (2008) Fatigue improvement of weld repaired crane runway girder by ultrasonic impact treatment. Weld World 52(11/12):50–62

6

CHAPTER

Characterisation of Local Grain Size Variation of Welded Structural Steel

Pauli Lehto, Jani Romanoff, Heikki Remes, Teemu Sarikka

Department of Mechanical Engineering, School of Engineering, Aalto University, P.O. Box 12200, FIN-00076 Aalto, Finland

ABSTRACT

Previously, it has been shown that the grain size distribution plays an important role in the mechanical properties of welded steel. In the previous investigation, the volume-weighted average grain size has been shown to capture the influence of grain size distribution, resulting in a better fitting Hall–Petch relationship between grain size and hardness. However, the previous studies exclude the effects arising from local variation in grain size. In this paper, the grain size measurement methods are extended for the characterisation of the local grain size variation, which is significant for welded joints and can have an adverse effect on mechanical properties. The local gradient of grain size variation and its dependency on measurement direction are considered. In addition, examples of grain size and hardness variation are shown for S355 base metal and two weld metals, and characteristic differences are highlighted and discussed. The coarse-grained areas of a heterogeneous microstructure are found to have lower hardness than fine-grained areas. However, the surrounding microstructure, i.e. local grain size gradient, has an influence on the measured hardness values.

Keywords (IIW Thesaurus) Grain size Measurement Structural steels Hardness Microstructure Weld metal

1 INTRODUCTION

Microstructural characterisation of engineering materials is a necessity for understanding the relationships between microstructural quantities and mechanical properties. Grain size is one of the fundamental microstructural quantities and correlates with several properties, such as hardness, stress–strain curve and fatigue strength [1–7]. Based on the work of Hall [8] and Petch [9], a relationship was found between grain size and the mechanical properties of steel. For yield strength, this relationship is:

$$\sigma = \sigma_0 + kd^{-1/2}, \tag{1}$$

where σ_0 is the lattice friction stress required to move individual dislocations, k is a material-dependent constant known as the Hall–Petch slope, and d is the average grain size [10]. As the Hall–Petch relationship is related to the measure of grain size, the correct definition of grain size is crucial.

The most commonly reported microstructural measure in literature is the average grain, even though there is a large variety of, e.g. ASTM grain size measurement methods, available [11]. Moreover, orientation imaging microscopy gives the operator a large degree of freedom for defining the measurement methodology and parameters such as step size, grain boundary misorientation criteria and filtering of the data [12–14]. For these reasons, an extensive interlaboratory round-robin [15] was carried out in order to define the grain size measurement methodology for the ISO 13067 [16] standard. In addition to measurement of average grain size, other material-specific factors such as differences in phase structure and grain size distribution need to be considered.

For heterogeneous materials, the grain size distribution is of particular interest since it has been shown to influence the mechanical properties [10, 17–22]. Improved grain size measurement methods are thus required to enhance the understanding between grain size distribution and mechanical properties [11]. Accurate description of the microstructural heterogeneity is also required for the mesoscale modelling of material behaviour [12]. Welds are an extreme case of heterogeneity since it is present both in macroscopic scale across the joint and in microscopic scale within a single zone; see Fig. 1. The grain size characterisation of heterogeneous weld metals was studied by Lehto et al. [23]. The grain size measurements revealed that structural steel weld metals exhibit a large variety of grain size distributions that are noticeably broader than those of the base metal. To capture the influence of grain size distribution, the volume-weighted grain size measurement was utilised. It was shown that the Hall–Petch relationship's dependence on grain size distribution is eliminated when then volume-weighted average grain size (d_v) is used:

$$\sigma = \sigma_0 + kd_v^{-1/2} = \sigma_0 + kd^{-1/2}\left(1 + f\frac{\Delta d}{d}\right), \quad (2)$$

where f is a constant describing the relation between average and volume-weighted average grain size and $\Delta d/d$ is the relative grain size dispersion. The equation is reduced to the original Hall–Petch Eq. (1) in the theoretical case that all grains are the same size. Furthermore, if samples with similar grain size dispersion are compared the original Hall–Petch, Eq (1) is applicable.

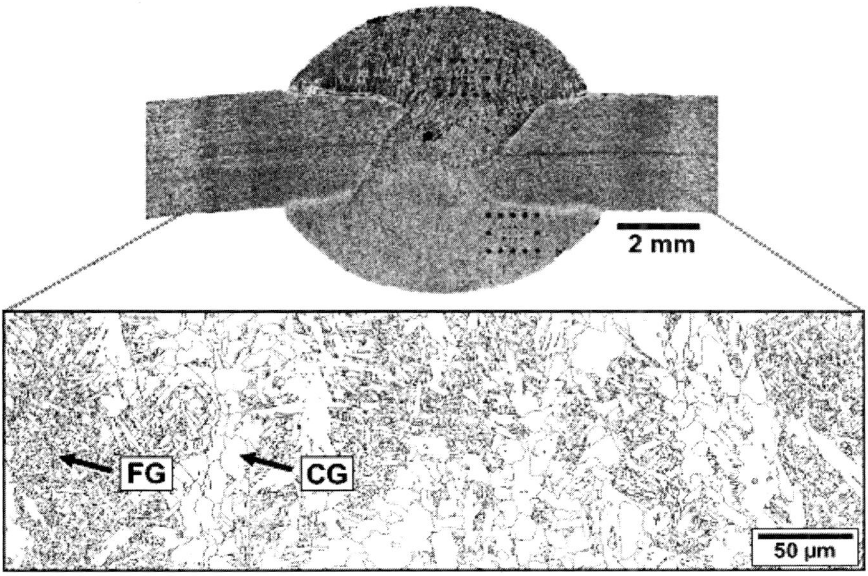

Figure. 1 Macro section of an arc welded joint (CV.1) and an example of weld metal grain size variation showing fine grained (FG) and coarse grained (CG) areas

While the previous work covered the microstructural characterisation of volume-weighted average grain size and its relation to the grain size distribution of welded structural steel, it did not provide further insight into the inherent large local variation of grain size. The aim of this paper is to characterise the local variation of grain size in welded joints and investigate how it influences the mechanical properties using hardness measurements. The point-sampled grain size measurement method is extended for the characterisation of local grain size variation both in numerical and visual form. The numerical approach is to present a moving average of grain size using line probes in different directions. In visual approach, the characterisation of grain size variation is further developed by substituting grain size (d) with the Hall–Petch grain size parameter ($d^{-0.5}$) in order to have a linear scale for grain size-dependent mechanical properties. The local variation of grain size is also compared to hardness measurement results.

2 MEASUREMENT OF LOCAL GRAIN SIZE VARIATION

Grain size is measured using the point-sampled intercept length method [24, 25]. The method is similar to the commonly used linear intercept method [26]; however, the measurements are carried out at random points instead of being measured along pre-determined lines. The method has previously been used for the characterisation of grain size distribution in welded structural steel [23]. Here, the method is extended for the characterisation of local grain size variation by including measurement direction-based averaging. In addition, moving averages of grain size are calculated across the microstructure. Figure 2 shows a flow chart of the measurement procedure using a fictional single-phase microstructure. Firstly, the grain size is measured for individual grains at random points that hit the grain interior (Fig. 2 (Ia)). The procedure is repeated a large number of times for one measurement direction, resulting in densely measured grain size for the individual grain (Fig. 2 (Ib)).

For an improved visual representation, the non-measured points are filled with a value from the nearest neighbouring point. As shown in (Fig. 2 (Ic)), the interpolation results in a good visual representation of grain size for the individual grain. Grain size is presented using the Hall–Petch grain size parameter ($d^{-0.5}$) in order to have a linear scale for grain size affected mechanical properties. This approach is taken to improve the resolution of visual representation in the grain size regime below 10 µm. Based on the Hall–Petch relationship, a small change of grain size in this regime has a significant effect on the mechanical properties, e.g. strength doubles as grain size decreases from 4 to 1 µm.; see Fig. 3. It is noted that the classical Hall–Petch relationship is applicable at grain sizes larger than 0.1 µm ($d^{-0.5} < 3.16$ µm$^{-0.5}$) [27, 28].

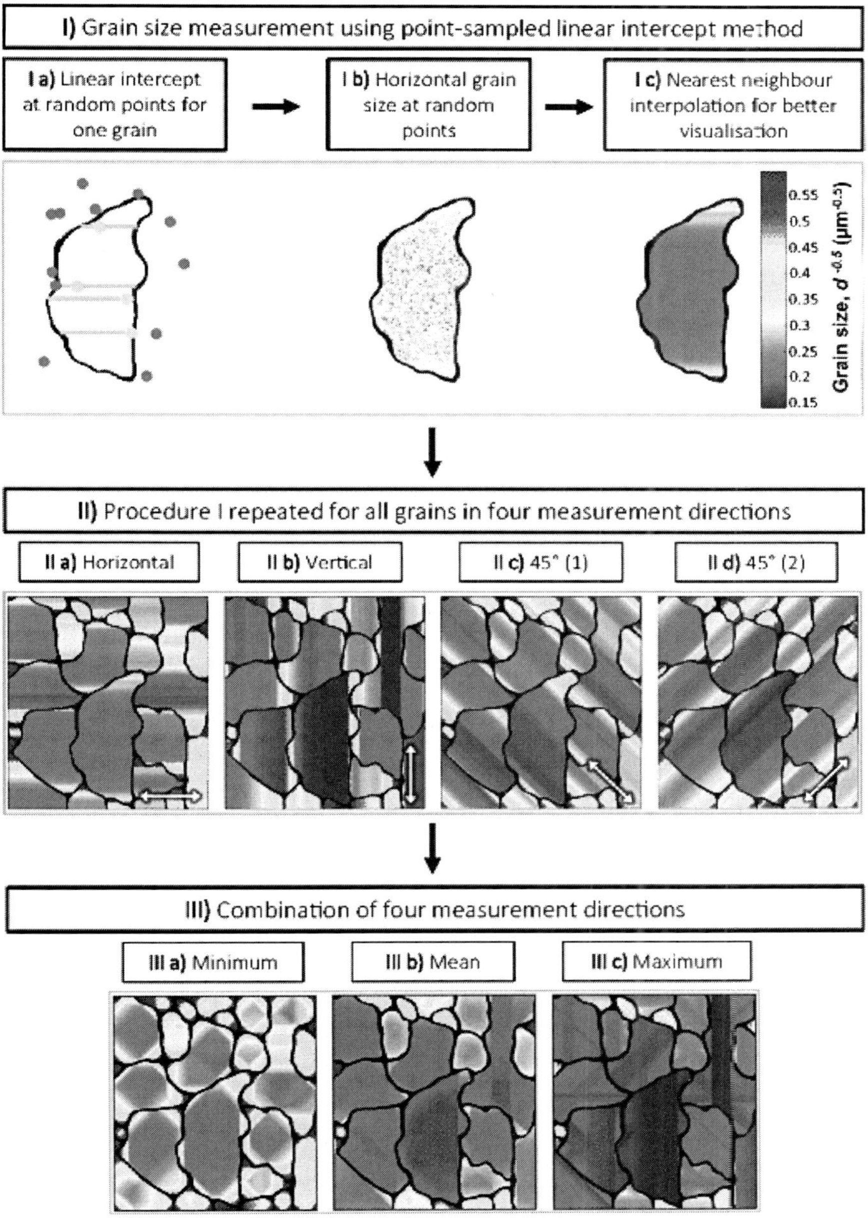

Figure. 2 Flowchart of the grain size measurement and analysis procedure

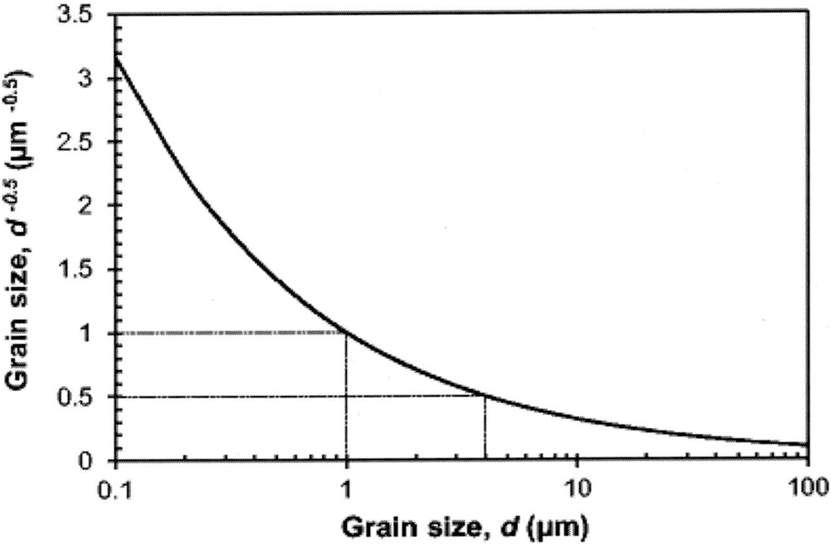

Figure. 3 Relationship between grain size (d) and the Hall–Petch grain size parameter ($d^{-0.5}$)

The procedures (Fig. 2 (Ia–c)) are repeated for all grains in the fictitious microstructure using four measurement directions (0°, 45°, 90°, and 135°) as shown in Fig. 2 (IIa–d). Typically, at least 25 % of the image points should be measured in each direction for accurate results. The probability P_i of a random point hitting a grain of size i is proportional to the surface area fraction of the grain. Based on relationships of stereology [29, 30], the surface area fraction provides a statistical estimator for the volume fraction:

$$P_i = \frac{A_i}{A_T} \approx \frac{V_i}{V_T}, \tag{3}$$

where A_i and V_i are the surface area and volume of a grain of size i. A_T and V_T are the total surface area and volume of all grains in the measurement domain, correspondingly. Thus, the measured grain size distribution can be considered as the volume-weighted grain size distribution if the assumption of isotropy is made. For further details and a Matlab implementation of the measurement procedure, the reader is referred to [23, 31].

To verify that the nearest neighbour interpolation does not introduce any bias or error to the data, the grain size distributions of the measured and interpolated data are compared in Fig. 4a. As the two counterparts overlap for each individual measurement direction, the interpolated data shown in (Fig. 2 (IIa–d)) can be used for further data analysis.

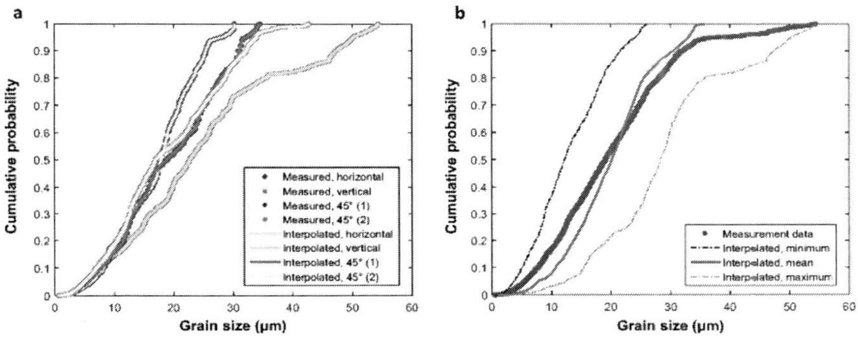

Figreu. 4 Cumulative probability distributions for grain size (**a**) measured and interpolated data for the four measurement directions (see Ib, Ic and II in Fig. 2). **b** Comparison of the analysed (see III in Fig. 2) and measurement data

In order to compare grain size with mechanical properties, e.g. hardness, the four measurement directions need to be combined into a single visualisation. The three alternatives used are to take the minimum, mean or maximum value of the four measurement directions at each point of all grains; see in Fig. 2 (IIIa–c). As shown in Fig. 4b, the grain size distributions of minimum and maximum cases are the lower and upper bounds for the measurement data, respectively. The mean has closely the same volume-weighted average grain size, d_v, as the measurement data even though the shape of the distribution is different. The agreement of volume-weighted average grain size has been verified for various heterogeneous microstructures, with the error typically being smaller than 1 %. Since the volume-weighted average grain size has been shown to correlate with hardness according to the Hall–Petch relationship [23], the mean grain size plot is used for further result analysis of hardness and grain size.

In addition to the above mentioned visual options, the moving averages of the minimum, mean and maximum grain size contours (Fig. 2 (IIIa–c)) are calculated across the micrograph using horizontal and vertical line probes. The line probes used in this study are 10-pixel wide. In addition, the border regions of the micrographs were excluded from the averaging to eliminate large grains that extend beyond the micrograph. Grain size at

90 % probability level was found as a suitable margin for exclusion at all edges of the image. Dimensions of the probes used are shown in Table 3. The difference between the three moving averages represents the local variation of grain size in different measurement directions.

3 EXPERIMENTS

3.1 Test specimens and sample preparation

The experiments are carried out for one base metal (BM.1) and two welded samples. The weld samples are flux-core arc welded either from two sides (CV.1) or from one side with a ceramic backing (CV.2). Plate edges were prepared by plasma cutting followed by grinding. The energy input for CV.1 is 3.5 kJ/cm for each side and for CV.2 it is 9 kJ/cm for the single weld bead. The macro sections of the weld samples CV.1 and CV.2 are shown in Figs. 1 and 5, correspondingly. All base metals are S355 grade ferritic-pearlitic structural steels; the base metal grades and mechanical properties are presented in Table 1.

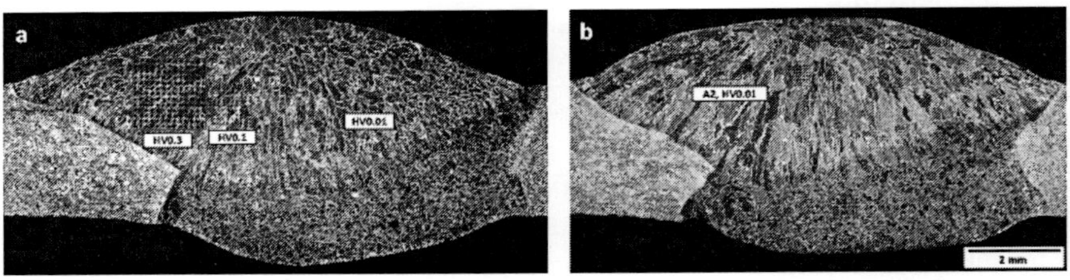

Figure. 5 Two macro sections of the arc welded joint CV.2 showing the hardness measurement locations of the heterogeneous weld metal presented in **a** Fig. 11 and **b** Fig. 12

Table 1 Test sample base metal grades and mechanical properties [23]

	Grade	Thickness (mm)	R_{EH} (MPa)	R_m (MPa)	A (%)
BM.1	GL D36	6	343	472	34
CV.1	S355J2	3	466	564	31.3
CV.2	S355J2	3	466	564	31.3

For this investigation, specimens are prepared from the base metal BM.1 and the two weld metals. To represent the extremities of the grain size distribution, two weld metal regions are chosen with homogeneous and heterogeneous grain size distributions. The root side weld metal of weld sample CV.1 is taken as the homogeneous weld metal, while the toe side weld metal of weld sample CV.2 is the heterogeneous weld metal. The three specimens are chosen since they have previously been found to closely follow the same relationship between grain size and hardness in the macroscopic scale using 70–100 μm hardness indentations [23]. The previously measured average hardness and grain size values, as well as phase volume fractions are presented in Table 2. The microstructural constituents were identified according to IIW document IX-1533-88 [32] by using the systematic manual point counting method according to ASTM E562-02 [33].

Table 2 Test specimen nomenclature, microstructural characterisation and the microstructural constituent volume fractions and the corresponding 95 % confidence intervals [23]

	Sample	Grain size	Hardness	Constituent volume fraction (%)		
		d_v (μm)	HM (MPa)	AF	PF	FC/P
Base metal	BM.1	15.29 ± 1.96	1412 ± 29	–	78.6 ± 5.3	21.4 ± 5.3 (P)
Homogeneous weld metal	CV.1 root	5.95 ± 0.20	1756 ± 17	7.3 ± 1.4	78.7 ± 4.7	14.0 ± 3.4
Heterogeneous weld metal	CV.2 toe	4.00 ± 0.45	1984 ± 34	37.5 ± 6.4	56.2 ± 5.3	6.3 ± 1.6

Abbreviations used: d_v volume-weighted average grain size, *HM* Martens hardness, *PF* primary ferrite, *AF* acicular ferrite, *FC* ferrite–carbide aggregate, *P* pearlite [23]

For the weld sample CV.2, an additional sample was prepared for EBSD analysis. The sample is used for comparing the grain size and hardness from the exact same location for the heterogeneous weld metal. This location is referred to as A2 in further instances. The location of this area and the hardness measurements carried out for the heterogeneous weld metal are shown in Fig. 5. All specimens were mounted in an electrically conductive resin and grinded up to P4000 grit abrasive paper. Polishing was done with 3 and 1 μm diamond paste and the additional heterogeneous sample with 0.25 μm as well. Before EBSD analysis, final polishing was carried out with colloidal silica in a vibratory polisher for 45 min.

3.2 **Material characterisation**

Instrumented indentation testing was used for measuring the mechanical properties. Hardness was measured with a CSM Instruments micro-indentation tester according to ISO 14577–1 [34] utilising large matrices containing up to 200 indentations. Hardness was defined using Martens hardness, denoted by *HM*, with a Vickers pyramid tip. The measurement data related to, but not presented in [23], indicates that *HM* correlates well with traditional Vickers hardness when presented in the same units; see Appendix 1 for further details. The test forces used equal HV0.01 (98.07 mN), HV0.1 (980.7 mN) and HV0.3 (2942.1 mN), producing indentations with approximately 7–10 μm, 30–35 μm and 50–60 μm diagonal lengths, respectively, depending on the sample. Indentation depth is approximately 1/7th of the indentation diagonal, i.e. 1–1.4 μm, 4.3–5 μm and 7.1–8.5 μm for the three measurement forces, respectively. Linear 30-s loading ramps were used, with a hold time of 10 s at the maximum force.

The heterogeneous microstructure was characterised using the electron backscatter diffraction (EBSD). EBSD analysis of the heterogeneous weld metal in sample CV.2 (location A2) was carried out prior to hardness measurements at a pre-marked area. A Zeiss Ultra 55 field emission scanning electron microscope equipped with a Nordlys F+ camera and Channel 5 software from Oxford Instruments was used for the EBSD analyses. The EBSD analyses were performed with a step size of 0.2 μm at a magnification of 1000× (area 373 × 280 μm). The acceleration voltage was 20 kV and the grain boundary misorientation criteria of 10° were used. Indexing rate of the EBSD maps was approximately 90 % and the EBSD data was post-processed using a nearest neighbour clean-up routine. The grain boundary maps from EBSD analysis were overlaid on optical micrographs and digital image processing applied in order to de-skew the distortions caused by stage drift during acquisition [35]. For optical micrographs, the samples were etched with a 2 % Nital solution. Digital image processing was carried out to create grain boundary maps from the optical micrographs for base metal and homogeneous weld metal; see [31] for more details.

4 **RESULTS**

4.1 **Microstructure and grain size**

The microstructures of the three specimens are shown in Fig. 6. The measured average grain sizes (d, d_v) of the single micrographs and the dimensions of the averaging line probes are presented in Table 3. The line probes are 10-pixel wide and thus the magnification of the micrograph affects the width of the probe. The relative grain size dispersion values for the three specimens are defined based on [23]:

$$\frac{\Delta d}{d} = \frac{d_{\max} - d_{\min}}{d} = \frac{d_{99\%} - d_{1\%}}{d}, \tag{4}$$

where the maximum and minimum grain sizes are taken as 99 and 1 % probability level grain sizes, respectively. This value characterises the grain size dispersion on a macroscopic scale for the entire micrograph, see Ref [23] for further details.

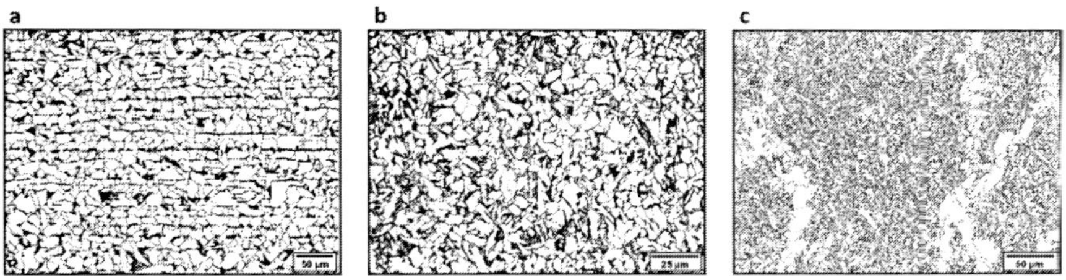

Figure. 6 The grain boundary maps used for grain size analysis of **a** base metal, **b** homogeneous weld metal and **c** heterogeneous weld metal

Table 3 Measured average grain sizes for the three microstructures presented in Fig. 6 and dimensions of the line probes used for grain size averaging

	Grain size			Line probe dimensions			Grains per line length	
	d (μm)	$\Delta d/d$ (−)	d_v (μm)	Line width (μm)	Horizontal (μm)	Vertical (μm)	Horizontal (−)	Vertical (−)
Base metal	10.18	3.16	15.43	1.84	532	385	35	25
Homogeneous weld	3.47	3.60	5.94	0.74	213	154	36	26
Heterogeneous weld	2.21	5.38	5.01	0.98	321	241	64	48

Grains per line length is defined as line length divided by the volume-weighted average grain size

The local grain size variation found in base metal as well as homogeneous and heterogeneous weld metals are compared in Fig. 7. Base metal and homogeneous weld metal have quite homogeneous grain size as is indicated by the relative grain size dispersion. The local variation of grain size is not significant and moreover the local variation is homogeneous throughout the microstructure. On the contrary, local variation of grain size is significant for the heterogeneous weld metal. The coarse-grained areas have inconsistent spacing as shown in Fig. 7c, and thus one grain boundary map of size 345 × 266 µm does not fully represent the length scale at which the coarse and fine-grained areas alternate.

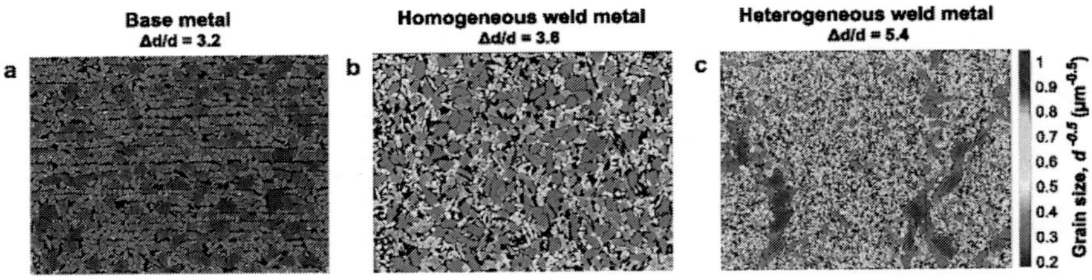

Figure. 7 Comparison of local grain size variation for S355 base metal, homogeneous and heterogeneous weld metals. The colour contours range between 0.18 and 1.05 µm$^{-0.5}$, representing the largest 99 % and smallest 1 % probability level grain sizes for the three specimens

4.2 **Characterisation of the local grain size variation**

The local grain size variation of base metal, homogeneous weld metal and heterogeneous weld metal is shown in Figs. 8, 9, and 10, correspondingly. The dimensions of the line probes used for the calculation of moving averages are presented in Table 3. It can be seen that the grain size of the base metal (Fig. 8) is very uniform when the mechanical properties are considered on grain scale according to the Hall–Petch relationship ($d^{-0.5}$). Variation around the volume-weighted average grain size (d_v) is minor for the moving average of the mean grain size. In terms of grain size, it is difficult to find strong or weak locations in the microstructure as the smallest and largest grains are within a narrow band of approximately 0.2–0.5 µm$^{-0.5}$. For homogeneous weld metal (Fig. 9), the variation of grain size is very similar to base metal, with little variation around the volume-weighted average grain size. Relatively, the minimum and maximum are further away from the mean curve, which is related to the broader grain size dispersion compared to base metal. This is visible as an increased amount of fine grains (acicular ferrite, *AF*) in between the coarse grains (primary ferrite, *PF*). For heterogeneous weld metal (Fig. 10), the location of coarse grains can be determined both visually and using the horizontal moving average of grain size. The areas with coarse grains are primary ferrite while fine-grained areas consist mostly of acicular ferrite. Furthermore, there is also noticeable grain size variation within acicular ferrite. Of the three specimens, the difference between minimum and maximum moving averages is largest for the heterogeneous weld metal in the area that consists primarily of acicular ferrite (X-coordinate 100–200 µm).

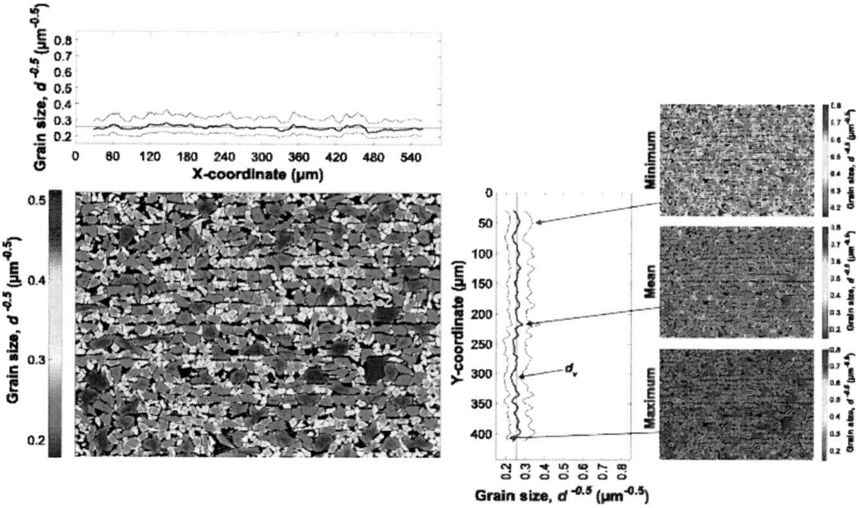

Figure. 8 Grain size plotted as a function of the Hall–Petch grain size parameter ($d^{-0.5}$) for the base metal (S355). The colour contour ranges from 99 to 1 % probability level grain size for the main figure. The colour contour range is extended for the side figures to cover the 1 and 99 % probability level grain sizes of the minimum and maximum cases

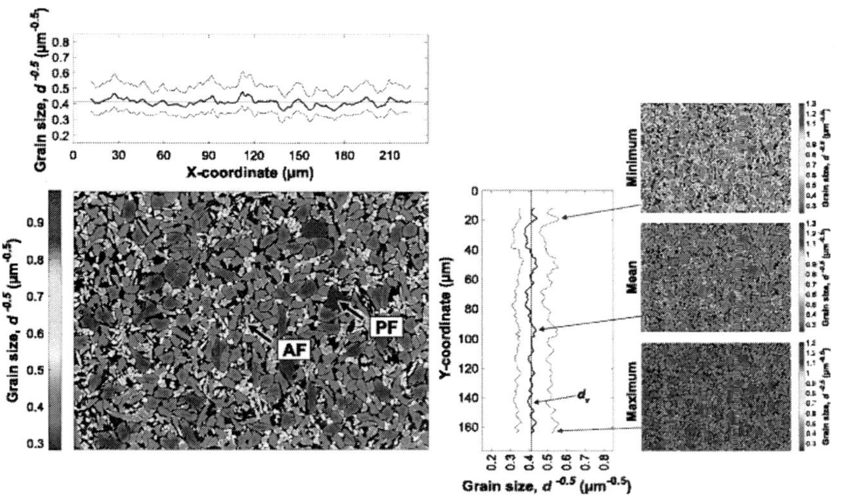

Figure. 9 Grain size plotted as a function of the Hall–Petch grain size parameter ($d^{-0.5}$) for the homogeneous weld metal. The colcur contour ranges from 99 to 1 % probability level grain size for the main figure. The colour contour range is extended for the side figures to cover the 1 and 99 % probability level grain sizes of the minimum and maximum cases

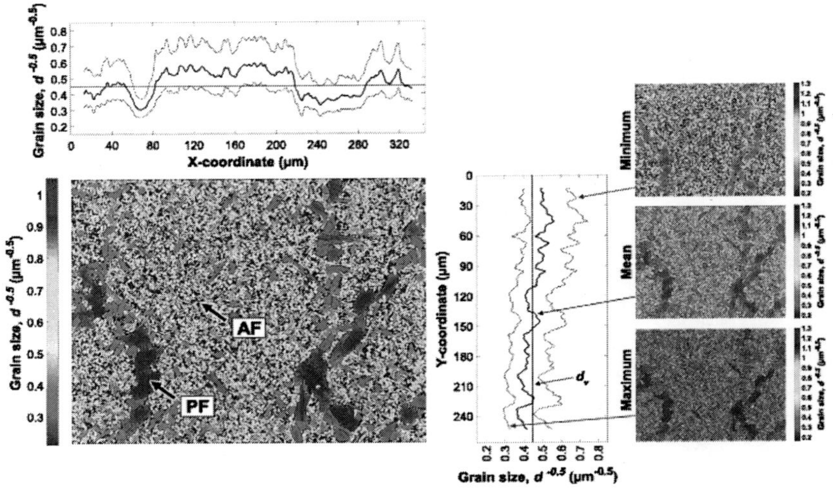

Figure. 10 Grain size plotted as a function of the Hall–Petch grain size parameter ($d^{-0.5}$) for the heterogeneous weld metal. The colour contour ranges from 99 to 1 % probability level grain size for the main figure. The colour contour range is extended for the side figures to cover the 1 and 99 % probability level grain sizes of the minimum and maximum cases

4.3 **Comparison of hardness variation in S355 base metal and weld metal**

The variation in grain size indicates that the mechanical properties might also vary significantly within the heterogeneous microstructure. The local variation in hardness measured using test forces HV0.01, HV0.1 and HV0.3 is shown in Fig. 11. The diagonal lengths of the indentations are approximately 7–10 µm, 30–35 µm and 50–60 µm, respectively. For clarity, the grain size comparison presented in Fig. 7 is from the same specimens but not at the exact location of the hardness measurements. Hardness is presented as an interpolated colour contour, and thus, it should be noted that the values in between the indentations do not represent real hardness values.

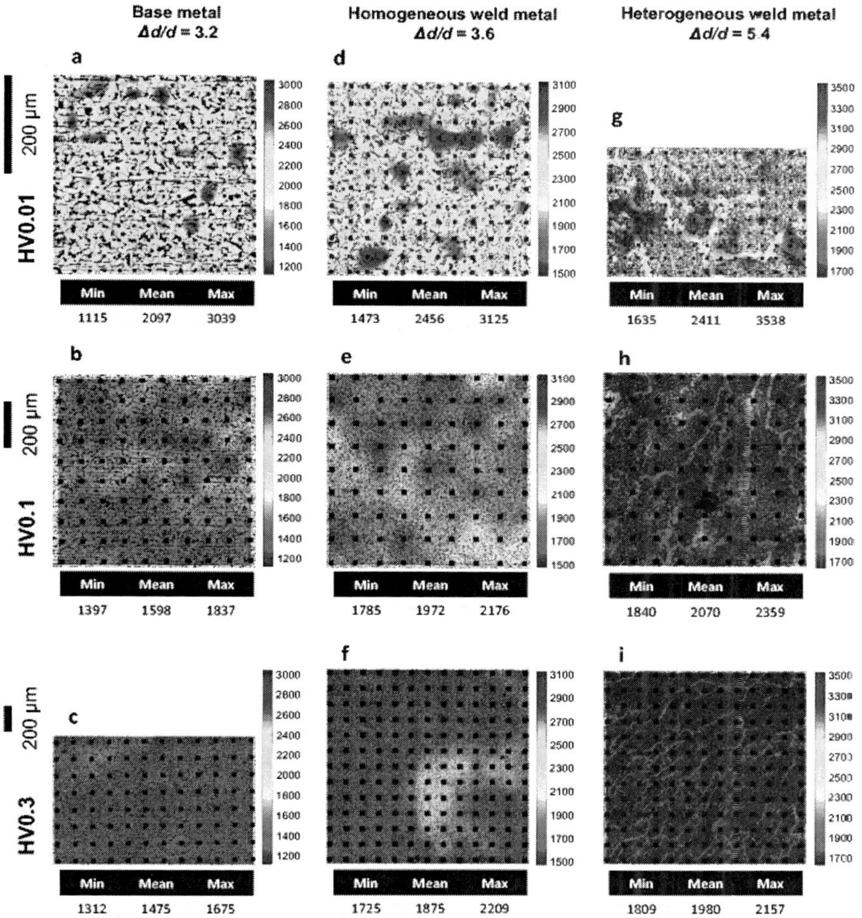

Figure. 11 Comparison of local hardness variation (*HM*) using three different indentation loads for the three specimens. The colour scales range between HV0.01 minimum and maximum for each specimen

For all specimens, it is observed that the mean hardness value increases with a decrease in indentation size. At the same time, the maximum hardness values increase significantly while minimum values remain approximately the same. Using the smallest test force of HV0.01, the difference between lowest and highest measured hardness values is significant, up to 0.9 times the average value. This can be related to the placement of the small indentations in the microstructure, hitting e.g. entirely in soft ferrite grain interiors or hard pearlite.

While base metal and homogeneous weld metal generally show smooth transitions between low and high hardness areas, the heterogeneous weld metal shows much higher local variation. This is the case particularly for the HV0.01 measurement force that produces small, 7–10 μm indentations. At higher test forces of HV0.1 and HV0.3, base metal and homogeneous weld metal show variation in hardness even though the grain size is very uniform in the structure. Likewise, with HV0.3 indentations, the local variation in hardness cannot be captured for the heterogeneous weld metal since the indentations are larger than the coarse-grained areas. Thus, it is challenging to choose indentation parameters; indentation size in relation to grain size and at the same time the spacing of indentations affects the data gathered from a pre-determined area. The transition and relations between different size indentations, as well as the relation to tensile properties require further study.

4.4 Correlation between grain size and hardness

To investigate the local grain size variation, additional hardness measurements were carried out for the heterogeneous weld metal. Prior to hardness measurements, an EBSD analysis was carried out at location A2 of the weld (see Fig. 5) that was pre-marked with hardness indentations. After hardness measurements, the specimen surface was etched, which enables overlaying of the EBSD maps on the microstructure and identifying the location of hardness indentations.

The hardness data and grain size contour for the studied heterogeneous weld metal are shown in Fig. 12. The figure contains (a) an optical micrograph of the hardness measurements, (b) visual estimate of correlation between hardness and local microstructure and (c) hardness in discrete form for individual indentations overlaid on the grain size contour. The visual estimate in (b) is divided into three categories: (1) green, grain size under the indentation correlates with hardness within approximately ±100 HM(52/96 samples, 54 %); (2) orange, hardness correlates with grain size when the microstructure surrounding the indentation is considered (18/96, 19 %) and (3) red, hardness and local grain size show no correlation (26/96, 27 %). Reader is referred to Appendix 2 for the hardness values and a figure that shows the grain size under and around each indentation.

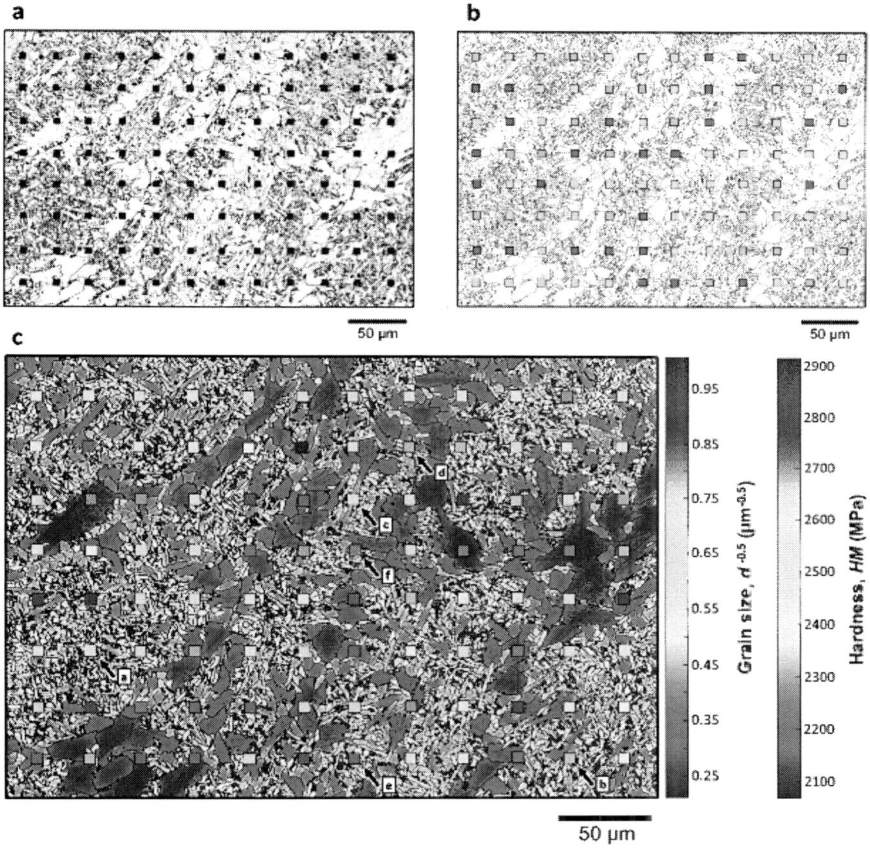

Figure. 12 a Optical micrograph of the hardness measurement area for the heterogeneous weld metal area A2, **b** grain boundary map and visual estimation of the correlation between local grain size and hardness; see text for the explanation and **c** the hardness value for each indentation overlaid on the grain size contour, ranging from 99 to 1 % probability level grain size

In general, the low hardness areas correspond with the coarse grains and high hardness with the fine-grained areas. A mixture of different grain sizes under the indentation is found to result in a measured hardness value that corresponds approximately with the average grain size of the area according to the rule of mixtures; see Fig. 13a. When fine grains are surrounded by coarse grains, hardness is found to decrease, and in the opposite case, to increase; see Fig. 13b, c, correspondingly. Placement of the indentation tip in a small cluster of fine grains is found to increase the hardness value, even though the microstructure under and around the indentation would otherwise consist of coarse grains; see Fig. 13d. In some cases, low hardness is measured in fine grains and high hardness in coarse grains; see Fig. 13e, f, correspondingly. Further numerical analysis of the microstructure at each indentation is required before correlations between local microstructure and hardness can be formulated.

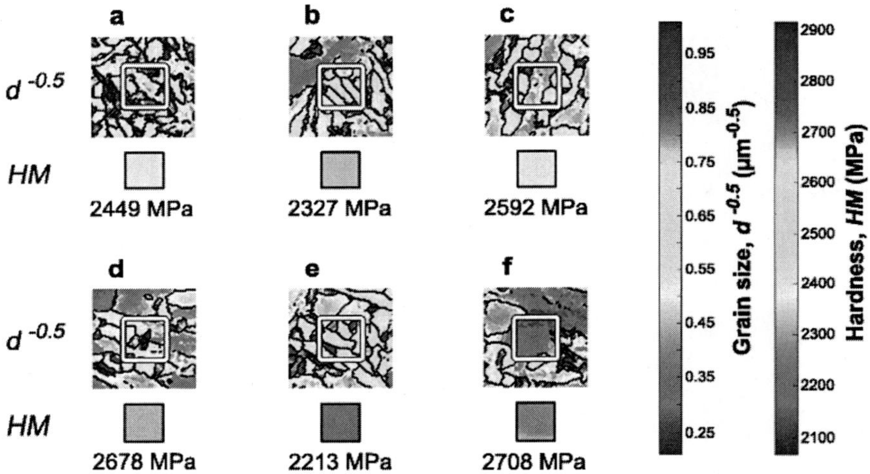

Figure. 13 Expanded view of grain size and hardness at indentations **a**–**f** indicated in Fig. 12c

5 DISCUSSION

The local grain size and hardness variation of ferritic base metal and two weld metals were studied. The point-sampled intercept length method was extended for the characterisation of local grain size variation. The local gradient of grain size variation and its dependency on measurement direction were considered. Base metal and homogeneous weld metal did not have significant local grain size variation, while the heterogeneous weld metal had distinct areas of coarse and fine grains. The coarse-grained areas are associated with primary ferrite and the fine-grained areas with acicular ferrite. Hardness was measured to investigate the influence of local grain size variation on mechanical properties.

EBSD grain size analysis is typically carried out using the grain identification and determination of grain size from its surface area, often taken as the diameter of a circle with an equivalent surface area [12, 36]. For homogeneous equiaxed microstructures, this assumption is justified, and values determined by circle equivalent diameter (d_{ceq}) and linear intercept are comparable within 10 % of each other [15]. The use of circle equivalent diameter has limitations when applied to heterogeneous weld metal microstructures. Welds can have microstructures where definition of individual grains is ambiguous due to discontinuities in the grain boundaries, and thus clusters of multiple grains can be detected as a single grain. This has been demonstrated for stainless steel pipe welds by Saukkonen et al. [37], also showing the robustness of the linear intercept method to EBSD indexing errors and the consequent grain detection. Furthermore, the circle equivalent diameter is not well suited for high aspect ratio grains by itself, and more information about the shape of the grain is required [38]. For welds, the high aspect ratio of grains, as well as morphological anisotropy, make the assumption of circular geometry invalid. Thus, the volume-weighted linear intercept method should be preferred for grain size measurement of heterogeneous weld microstructures with complex grain morphologies. To that end, the methodology presented here is able to consider the aspect ratio of grains through the four measurement directions. Although limited to four measurement directions, it seems sufficient for the characterization of weld metal microstructures. It is also not sensitive to small discontinuities in the grain boundaries that are observed in welded [39, 40] and deformed [41] steel microstructures. For a physically based definition, the grain size measurement direction should be parallel with the direction of the slip plane in each grain, requiring the utilisation of orientation information in grain size measurement. As the present approach is based on measurement of 2D sections, the morphological anisotropy of the grains is not considered.

Grain size variation was found to be significant within acicular ferrite in the heterogeneous weld metal, having the biggest relative difference between the minimum and maximum grain size curves in Fig. 10. The variation is related to the high aspect ratio of the grains, resulting in a significant difference between the shortest and longest grain size measurement directions. Local variation of grain size is visually significant within acicular ferrite, as shown in Figs. 10 and 12c, ranging approximately between 1 and 5 μm even though not revealed by the moving average line probe. The line probes are influenced by the number of grains falling under the line, and thus, the true variation will average out with a large number of grains. This is the case for acicular ferrite in the heterogeneous weld metal, and further research is required to define methods better suited for quantifying the local grain size variation numerically.

Even though the base metal and homogeneous weld metal are uniform in grain size, hardness measurements revealed low and high hardness regions using HV0.1 and HV0.3 test forces; see Fig. 11b, c, e and f. Visually, the grain size is uniform through the entire microstructure, and thus, it seems that factors other than grain size contributed to the observed differences. Hardness measurements using HV0.01 test force revealed that in the heterogeneous weld metal, the low hardness values are usually associated with coarse grains and high hardness values with fine grains, as is expected by the Hall–Petch relationship [8, 9]; see Fig. 12c. Despite the visual estimate of 54 % of hardness measurements corresponding with the local grain size (Fig. 12b), variation is observed between regions of similar grain size. This can be caused by differences in grain orientation [42, 43] and dislocation slip transmission between grains [44, 45].

The plastic zone from a hardness indentation is hemispheric, extending 1.0–1.9 times the contact radius with the highest plastic strain (>20 %) located directly underneath the indenter contact region [46]. Therefore, the microstructure of the entire plastic zone will influence the hardness measurement. This is visible for the 19 % of measurements where surrounding coarse or fine grains either decreased or increased the hardness value, respectively. The limitation of observing grain size on the surface of the specimen is visible with 27 % of the measurements as the grain size is not in agreement with the measured hardness value. Furthermore, the microstructural length scale, e.g. grain size, interacts with the length scale of hardness measurements [47]. These effects become relevant with indentations small relative to the grain size, and disappear as the length scale of deformation is large relative to the microstructural heterogeneities [47]. For the heterogeneous weld metal, the coarse grains are up to 25 μm in size while the fine grains are 1–5 μm in size, and thus the plastic zone of an indentation with 7-μm diagonal and 1-μm depth can be limited to a few grains or tens of grains. Therefore, the size of the plastic zone can vary depending on grain size and should be considered in the grain size measurement.

In addition to grain size characterisation, phase properties need to be considered in microstructural analysis. Even though in case of acicular ferrite and primary ferrite the phases behave similarly in hardness measurements [23], the tensile properties and toughness can show significant differences despite similarity of grain size. Zhao [48] found that acicular ferrite ($d = 4$–5 μm) had lower ultimate tensile strength but higher yield strength than ultrafine-grained ferrite ($d = 1$ μm). The differences were attributed to carbonitride precipitation, higher dislocation density and the lath bundle size of acicular ferrite. In general, acicular ferrite is beneficial for mechanical properties and besides tensile properties it improves impact toughness and decreases the transition temperature [49, 50]. Thus, methods are sought after to increase the volume fraction of acicular ferrite in welds [49, 51].

In this study, the weld metal microstructures were examined using two dimensional analysis. Based on current analysis hardness shows good agreement (54 %) in locations where the local grain size gradient is typically small. It is expected that the gradient of grain size under the indentation has a significant effect on measured hardness values as well. It is likely that for the locations with good agreement the grain size gradient under the indentation is small or that the gradient is such that the measured hardness appears to correlate with surface grain size. For weld metals, the solidification behaviour controls the grain structure and the achieved mechanical properties [52]. The competitive growth of grains, influenced by the transient thermal conditions and solidification characteristics of the weld metal, determines the morphology of the grain structure [52, 53]. In fusion welding axial, columnar and equiaxed grain morphologies can be observed depending on the thermal gradient, cooling rate and use of grain refining particles [52, 54, 55]. Therefore, the appearance of the grains on one cross section may not be representative of the grain morphology [56], causing bias for the assumed relationship between surface area and volume (Eq. 3). For example, columnar grains can appear very large on the transverse section even though the grains are quite shallow. Thus, the morphology of the grains and the anisotropy of the microstructure in the direction of the weld bead should be included in the analysis. It is expected that hardness of the heterogeneous weld metal (Fig. 12) is better predicted if the morphological anisotropy is considered. Furthermore, other obstacles to dislocation motion such as inclusions can influence the measured hardness values on a local scale. Consideration of these aspects is left for future work.

6 CONCLUSIONS

The point-sampled grain size measurement method was extended to the characterisation of local grain size variation. The Hall–Petch grain size parameter ($d^{-0.5}$) was found to give a good visual representation of grain size-dependent mechanical properties. Heterogeneous weld metal was found to have significant local variation of grain size while base metal and homogeneous weld metal were not. Furthermore, the local variation of grain size correlates with hardness measurements for a large portion of the measurements, with coarse grains generally showing low hardness and fine grains high hardness. Grain size alone was not able to explain all of the measurement results and thus future work is needed to analyse the local microstructure to determine factors other than grain size that should be included in the microstructural characterisation. In particular, the morphological anisotropy, i.e. the three dimensional shape of the grains needs to be characterized.

ACKNOWLEDGMENTS

The work has been done within the FIMECC BSA (Breakthrough Steels and Applications) programme as part of the FIMECC Breakthrough Materials Doctoral School. We gratefully acknowledge the financial support from the Finnish Funding Agency for Innovation (Tekes) and the participating companies. Funding from the Academy of Finland project "Fatigue of Steel Sandwich Panels" (FASA) under grant agreement no. 261286 is gratefully appreciated. The constructive comments of Professor Sven Bossuyt from Aalto University School of Engineering are gratefully acknowledged. Tuomo Nyyssönen from Tampere University of Technology is acknowledged for providing EBSD data suitable for the preliminary development of the grain size measurement Matlab code.

APPENDIX 1: VICKERS HARDNESS COMPARED TO INSTRUMENTED INDENTATION HARDNESS

Traditional hardness measurements rely on measurement of, e.g. the diagonal of the indentation after the test force has been removed. As a result, only the residual plastic deformation is considered and elastic deformation is ignored. Instrumented indentation testing (IIT) enables the evaluation of both the elastic and plastic deformation by monitoring of the test force and displacement of the indenter. In addition to hardness, other material parameters such as indentation modulus can be determined without the need for optical measurement of the indentation [34].

Due to the force-displacement data being available, several definitions can be used for defining hardness from IIT measurements, with indentation hardness (HIT) being the most commonly used parameter. However, the definition of indentation hardness [34] is the maximum force divided by the projected area, which is not the case for definition of traditional Vickers hardness. ISO 14577–1 Annex F [34] includes the correlation of HIT to Vickers hardness by converting the projected area to the surface area of contact. The correlation between the two is formulated as $HV = 0.0945\ x\ HIT$ based on the constant ratio of projected area to surface area and unit conversion to kg/mm². It is noted that the values calculated in this manner should not be used as a substitute for Vickers hardness.

Another approach is to use Martens hardness (HM), which by definition is the same as traditional Vickers hardness. These two options are compared in Fig. 14 for the measurements of Ref. [23], although not presented there in this form. For comparability, Martens hardness is also converted to kilogrammes per millimetre squared. The ferritic weld metals (WM) have good correlation with Martens hardness and Vickers hardness below 250 HV within the ±5 % limits. At higher values, Martens hardness values are lower than Vickers hardness, indicating that significant in-plane elastic recovery took place after the test force was removed, thus reducing the optically measured diagonals. Indentation hardness (HIT) shows higher deviation at low Vickers hardness values. The values are consistently higher than Vickers hardness, following approximately the +5 % line above 200 HV. Due to the good agreement of Martens hardness below 250 HV, it is used for the ferritic samples in the current study.

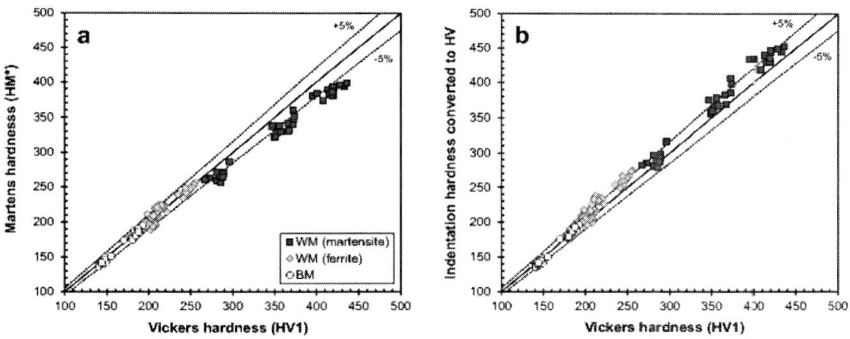

Fig. 14 Comparison of base metal (BM) and weld metal (WM) Vickers hardness (HV1) to **a**Martens hardness (HM) and **b** indentation hardness converted to Vickers hardness. The*asterisk* indicates conversion of units from MPa to kg/mm². The measurements are from Ref. [23], although not presented there in this form

APPENDIX 2: GRAIN SIZE CONTOUR AND HARDNESS VALUES

This appendix provides an alternative representation of the grain size contour presented in Fig. 12. In order to reveal the microstructure under each indentation, the hardness indentations have been moved below each indentation in Fig. 15. The measured hardness values are presented in Table 4.

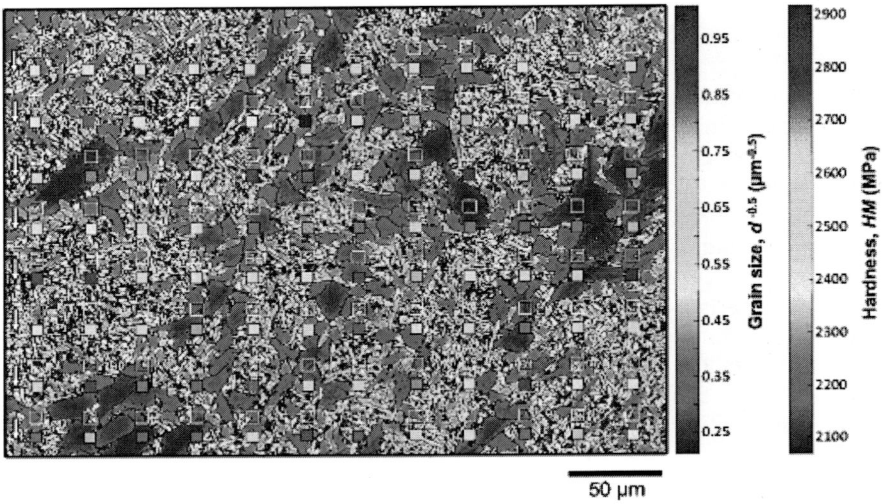

Fig. 15 The hardness value for each indentation overlaid on mean grain size contour for the heterogeneous weld metal area A2. Hardness values are shown below each indentation, indicated by the *arrows* on the left side, to reveal the grain size under the indentations. The grain size colour contour ranges from 99 to 1 % probability level grain size

Table 4 Martens hardness (*HM*) values [MPa] for the indentations presented in Figs. 12 and 15

		Indentation number											
		1	2	3	4	5	6	7	8	9	10	11	12
Indentation row	8	2615	2587	2548	2585	2515	2549	2464	2557	2629	2593	2283	2485
	7	2605	2726	2609	2437	2396	2068	2356	2678	2320	2514	2445	2530
	6	2657	2701	2295	2559	2737	2192	2592	2522	2751	2332	2549	2312
	5	2564	2405	2526	2484	2239	2691	2708	2488	2259	2238	2259	2323
	4	2747	2763	2554	2540	2494	2432	2145	2322	2411	2449	2523	2113
	3	2641	2449	2513	2309	2441	2524	2170	2361	2468	2256	2491	2586
	2	2341	2743	2258	2200	2720	2521	2513	2292	2421	2133	2360	2402
	1	2174	2321	2200	2197	2349	2143	2213	2453	2596	2247	2327	2396

Indentation (1, 1) is located in the lower left corner and indentation (12, 8) in the upper right corner of the figures

Indentation (1, 1) is located in the lower left corner and indentation (12, 8) in the upper right corner of the figures

REFERENCES

1 Hall EO (1954) Variation of hardness of metals with grain size. Nature 173:948–9

2 Armstrong RW, Codd I, Douthwaite RM, Petch NJ (1962) The plastic deformation of polycrystalline aggregates. Philos Mag 7:45–58

3 Armstrong RW (1970) The influence of polycrystal grain size on several mechanical properties of materials. Metall Mater Trans 1:1169–76

4 Tachibana S, Kawachi S, Yamada K, Kunio T (1988) Effect of grain refinement on the endurance limit of plain carbon steels at various strength levels. Nippon Kikai Gakkai Ronbunshu, A Hen/Transactions Japan Soc Mech Eng Part A 54:1956–61

5 Furukawa M, Horita Z, Nemoto M, Valiev RZ, Langdon TG (1996) Microhardness measurements and the Hall–Petch relationship in an Al-Mg alloy with submicrometer grain size. Acta Mater 44:4619–29.

6 Chapetti M, Miyata H, Tagawa T, Miyata T, Fujioka M (2004) Fatigue strength of ultra-fine grained steels. Mater Sci Eng A 381:331–6.

7 Hansen N (2004) Hall–Petch relation and boundary strengthening. Scr Mater 51:801–6.

8 Hall EO (1951) The deformation and ageing of mild steel: III discussion of results. Proc Phys Soc Sect B 64:747–53

9 Petch NJ (1953) The cleavage strength of polycrystals. J Iron Steel Inst 174:25–8

10 Masumura RA, Hazzledine PM, Pande CS (1998) Yield stress of fine grained materials. Acta Mater 46:4527–34.

11 Roebuck B (2000) Measurement of grain size and size distribution in engineering materials. Mater Sci Technol 16:1167–74CrossRef

12 Mingard KP, Roebuck B, Quested P, Bennett EG (2010) Challenges in microstructural metrology for advanced engineered materials. Metrologia 47:S67–82.

13 Mingard KP, Roebuck B, Bennett EG, Gee MG, Nordenstrom H, Sweetman G et al (2009) Comparison of EBSD and conventional methods of grain size measurement of hardmetals. Int J Refract Met Hard Mater 27:213–23.

14 Mingard KP, Day AP, Quested PN (2014) Recent developments in two fundamental aspects of electron backscatter diffraction. IOP Conf Ser Mater Sci Eng 55:012011.

15 Mingard KP, Quested PN, Peck MS (2012) Determination of grain size by EBSD - Report on a round robin measurement of equiaxed Titanium

16 ISO (2012) ISO 13067 - Microbeam analysis. Electron backscatter diffraction. Measurement of average grain size

17 Kurzydlowski KJ, Bucki JJ (1993) Flow stress dependence on the distribution of grain size in polycrystals. Acta Metall Mater 41:3141–6

18 Weertman JR, Sanders PG, Youngdahl CJ (1997) The strength of nanocrystalline metals with and without flaws. Mater Sci Eng A 234–236:77–82

19 Morita T, Mitra R, Weertman JR (2004) Micromechanics model concerning yield behavior of nanocrystalline materials. Mater Trans 45:502–8.

20 Berbenni S, Favier V, Berveiller M (2007) Micro–macro modelling of the effects of the grain size distribution on the plastic flow stress of heterogeneous materials. Comput Mater Sci 39:96–105.

21 Raeisinia B, Sinclair CW, Poole WJ, Tomé CN (2008) On the impact of grain size distribution on the plastic behaviour of polycrystalline metals. Model Simul Mater Sci Eng 16:025001.

22 Ramtani S, Bui HQ, Dirras G (2009) A revisited generalized self-consistent polycrystal model following an incremental small strain formulation and including grain-size distribution effect. Int J Eng Sci 47:537–53.

23 Lehto P, Remes H, Saukkonen T, Hänninen H, Romanoff J (2014) Influence of grain size distribution on the Hall–Petch relationship of welded structural steel. Mater Sci Eng A 592:28–39.

24 Gundersen HJG, Jensen EB (1983) Particle sizes and their distributions estimated from line- and point-sampled intercepts. Including graphical unfolding. J Microsc 131:291–310

25 Gundersen HJG, Jensen EB (1985) Stereological estimation of the volume-weighted mean volume of arbitrary particles observed on random sections. J Microsc 138:127–42

26 ASTM E1382 - 97 (2004) Standard test methods for determining average grain size using semiautomatic and automatic image analysis. ASTM International, West Conshohocken.

27 Takeuchi S (2001) The mechanism of the inverse Hall–Petch relation of nanocrystals. Scr Mater 44:1483–7.

28 Fan G, Choo H, Liaw P, Lavernia E (2005) A model for the inverse Hall–Petch relation of nanocrystalline materials. Mater Sci Eng A 409:243–8.

29 Underwood EE (1970) Quantitative stereology. Addison-Wesley Publishing Co., Reading

30 ASTM E1245 - 03 (2003) Standard practice for determining the inclusion or second-phase constituent content of metals by automatic image analysis. ASTM International, West Conshohocken.

31 Lehto P (2015) Aalto University wiki - grain size measurement using Matlab.https://wiki.aalto.fi/display/GSMUM

32 (1991) Guide to the light microscope examination of ferritic steel weld metals, Doc. IIW IX-1533-88. Weld World 29:160–76

33 ASTM E 562 – 02 (2002) Standard test method for determining volume fraction by systematic manual point count. ASTM International, West Conshohocken.

34 ISO 14577–1 (2002) Metallic materials. Instrumented indentation test for hardness and materials parameters. Part 1: Test method. International Organization for Standardization, Geneva

35 Gee M, Mingard K, Roebuck B (2009) Application of EBSD to the evaluation of plastic deformation in the mechanical testing of WC/Co hardmetal. Int J Refract Met Hard Mater 27:300–12.

36 Mingard KP, Roebuck B, Bennett EG, Thomas M, Wynne BP, Palmiere EJ (2007) Grain size measurement by EBSD in complex hot deformed metal alloy microstructures. J Microsc 227:298–308.

37 Saukkonen T, Aalto M, Virkkunen I, Ehrnstén U, Hänninen H (2011) Plastic strain and residual stress distributions in an AISI 304 stainless steel BWR pipe weld. 15th Int. Conf. Environ. Degrad, p. 2351–67

38 Radwański K, Wrożyna A, Kuziak R (2015) Role of the advanced microstructures characterization in modeling of mechanical properties of AHSS steels. Mater Sci Eng A 639:567–74.

39 Nie W, Shang C, You Y, Zhang X, Subramanian S (2013) Microstructure and toughness of the simulated welding heat affected zone in X100 pipeline steel with high deformation resistance. Acta Metall Sin 48:797–806.

40 Sabooni S, Karimzadeh F, Enayati MH, Ngan a HW (2015) Friction-stir welding of ultrafine grained austenitic 304L stainless steel produced by martensitic thermomechanical processing. Mater Des 76:130–40.

41 Gazder A, Cao W, Davies CHJ, Pereloma EV (2008) An EBSD investigation of interstitial-free steel subjected to equal channel angular extrusion. Mater Sci Eng A 497:341–52.

42 Haušild P, Materna A, Nohava J (2014) Characterization of anisotropy in hardness and indentation modulus by nanoindentation. Metallogr Microstruct Anal 3:5–10.

43 Stinville JC, Tromas C, Villechaise P, Templier C (2011) Anisotropy changes in hardness and indentation modulus induced by plasma nitriding of 316L polycrystalline stainless steel. Scr Mater 64:37–40.

44 Patriarca L, Abuzaid W, Sehitoglu H, Maier HJ (2013) Slip transmission in bcc FeCr polycrystal. Mater Sci Eng A 588:308–17.

45 Soer W, De Hosson JTM (2005) Detection of grain-boundary resistance to slip transfer using nanoindentation. Mater Lett 59:3192–5.

46 Durst K, Backes B, Göken M (2005) Indentation size effect in metallic materials: correcting for the size of the plastic zone. Scr Mater 52:1093–7.

47 Lilleodden E, Nix W (2006) Microstructural length-scale effects in the nanoindentation behavior of thin gold films. Acta Mater 54:1583–93.

48 Zhao MC, Yang K, Shan YY (2003) Comparison on strength and toughness behaviors of microalloyed pipeline steels with acicular ferrite and ultrafine ferrite. Mater Lett 57:1496–500.

49 Fattahi M, Nabhani N, Hosseini M, Arabian N, Rahimi E (2013) Effect of Ti-containing inclusions on the nucleation of acicular ferrite and mechanical properties of multipass weld metals. Micron 45:107–14.

50 Seo JS, Lee C, Kim HJ (2013) Influence of oxygen content on microstructure and inclusion characteristics of bainitic weld metals. ISIJ Int 53:279–85

51 Seo JS, Kim HJ, Lee C (2013) Effect of Ti addition on weld microstructure and inclusion characteristics of bainitic GMA welds. ISIJ Int 53:880–6.

52 Han R, Lu S, Dong W, Li D, Li Y (2015) The morphological evolution of the axial structure and the curved columnar grain in the weld. J Cryst Growth 431:49–59.

53 Han R, Dong W, Lu S, Li D, Li Y (2014) Modeling of morphological evolution of columnar dendritic grains in the molten pool of gas tungsten arc welding. Comput Mater Sci 95:351–61.

54 Kidess A, Tong M, Duggan G, Browne DJ, Kenjeres S, Richardson I et al (2015) An integrated model for the post-solidification shape and grain morphology of fusion welds. Int J Heat Mass Transf 85:667–78.

55 Kou S, Le Y (1986) Nucleation mechanism and grain refining of weld metal. Weld J 65:305–13

56 Tan W, Shin YC (2015) Multi-scale modeling of solidification and microstructure development in laser keyhole welding process for austenitic stainless steel. Comput Mater Sci 98:446–58.

7

CHAPTER

Dissimilar Metal Joining of Zinc Coated Steel and Aluminum Alloy by Laser Roll Welding

Hitoshi Ozaki[1] and Muneharu Kutsuna[2]

[1] Graduate School of Engineering, Mie University,, Japan

[2] Advanced Laser Technology Research Center Co., Ltd.,, Japan

1. INTRODUCTION

Nowadays, the car industry has targets to improve fuel consumption due to the problems of the global environment. For example, in the Corporate Average Fuel Economy, CAFE, which was approved in the U. S. Congress in July 2011, attaining the average fuel efficiency of 54.5 mpg (23.2 km/l) by 2025 is called for (Yamamoto, 2012). In addition, it's obliged to improve the average fuel efficiency to 35.5 mpg (15.1 km/l) by 2017. This is a significant increase compared with the present 27.5 mpg (11.7 km/l) which have been fixed for the past 11 years. Hence, the further weight reduction of cars has become imperative.

On the other hand, the weight of vehicle which influences the fuel consumption directly is increasing owing to a rise of safety awareness of automobile user's, tightening of safety standards, and diversification and sophistication of needs. Therefore, the car industry has conflicting targets of the low fuel consumption by "lightening car body" and "safety improvement". Additionally these targets should be achieved in together. Then, a "hybrid body structure" concept has been proposed.

In this concept, high strength steel and light alloy are arranged in the right places of the car body. For instance, by using high strength steel for pillars, a lightening with improving strength is achieved. In addition, by using aluminum alloys for the bonnet, the door panel, and the trunk lid, a further lightening is achieved. In order to make this "hybrid body structure", the welding technology for joining of steel to aluminum with high reliability and productivity is required. However, it's difficult to join steel to aluminum by conventional fusion welding.

Generally, in respect to dissimilar metal joining by the fusion welding, because of melting and alloying largely both metals, intermetallic compound layer formed at joint interface grows thick; oxide firm is formed; hot cracking is generated. Consequently, high joint strength can not be gotten. Moreover, when melting point of both metals has large difference, burn-through occurs in the metal of lower melting point (Nishimoto et al., 2005).

In order to solve these problems, many studies of the dissimilar metal joining have been conducted such as resistance welding (Okita, 2004), laser welding (Katayama, 2004), explosive welding (Satou, 2004), friction stir welding (Okamura & Aota, 2004; Katoh & Tokisue, 2004), and diffusion bonding (Ohashi, 2004). As a result, solid-phase bonding is mainly put to practical use. However, higher productivity, joint strength and flexibility are needed to expand the coverage of application of the dissimilar metal joints. Since laser welding has advantages such as local heat input, short process time and high flexibility compared with other welding processes, there has been considerable research on the welding of steel and aluminum. In the recent, the studies of keyhole welding (Torkamany, et al., 2010), laser welding-brazing (Dharmendra, et al., 2011), dual-beam YAG laser welding (Yan, et al., 2010), and thermal cycle during laser welding (Fan, et al., 2011) are carried out.

Then, the welding process of steel and aluminum which have been regarded as difficult, Laser Roll Welding has been developed for joining of dissimilar metals by M. Kutsuna, M. Rathod and A. Tsuboi in 2002. This welding is a hybrid welding process, as shown in Fig.1, combined high-temperature heating by a 2.4 kW CO_2 laser with pressurizing by a pressure roller. It's registered as Japanese Patents No. 3535152 and No. 3692135.

Fig.2 shows Fe-Al phase diagram (Massalski, et al., 1986). In this figure, there are varioius intermetallic compounds, hereafter, called as IMC's, and they are grouped as Fe-rich compounds, FeAl and Fe_3Al, and Al-rich compounds, $FeAl_2$, Fe_2Al_5 and $FeAl_3$. M. Yasuyama et al. (1996) have shown the mechanical properties of these IMC's. Vickers hardness of cast IMC's is shown in Table 1, and mechanical properties by compressive test are shown in Fig.3.

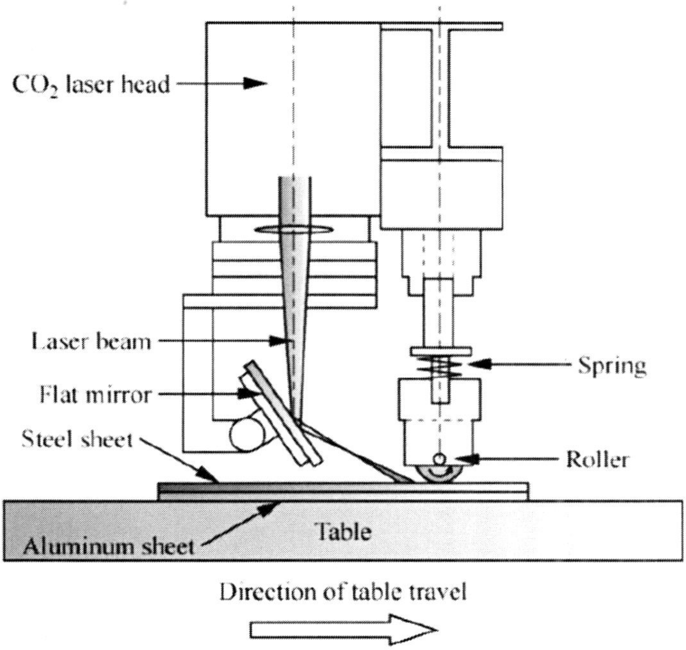

FIGURE 1. Schematic diagram of Laser Roll Welding process

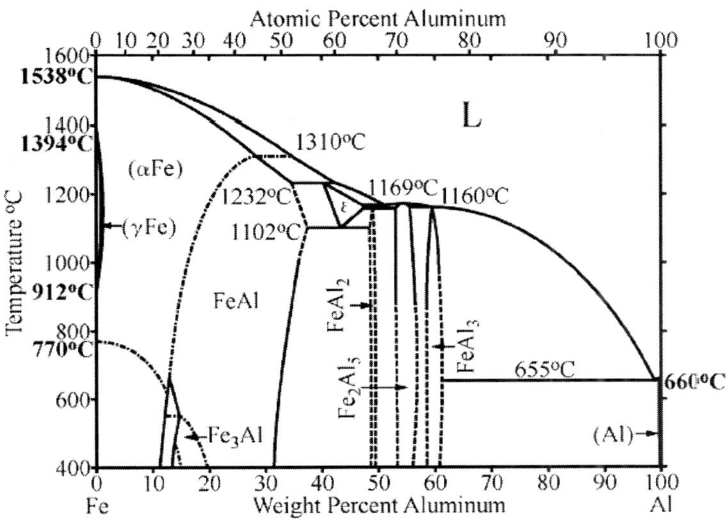

FIGURE 2. Fe-Al binary equilibrium diagram

TABLE 1. Vickers hardness of Fe-Al intermetallic compounds

	Vickers hardness
$FeAl_3$	892
Fe_2Al_5	1013
FeAl	470
Fe_3Al	330

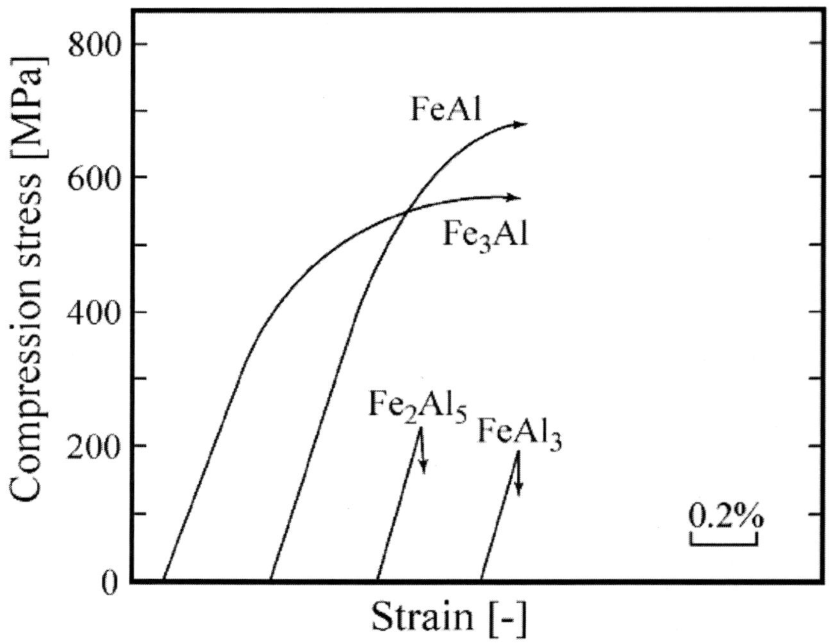

FIGURE 3. Stress-strain curves in compressive test of Fe-Al intermetallic compounds

While Al-rich IMC's are hard and brittle, Fe-rich IMC's show slight ductility and high strength. These brittle IMC's constitute barriers to the dissimilar metal welding of steel to aluminum. Kutsuna et al. thought if the thickness of brittle IMC layer is minimized and the formation of more ductile IMC's is promoted, high reliable joints are able to be obtained. Therefore, they developed the Laser Roll Welding process. In this process, since the thermal cycle for joining can be shortened by laser heating, the formation of the brittle IMC's can be easily controlled. Furthermore, good contact of a steel sheet and an aluminum sheet and rapid heat transfer from the steel sheet to the aluminum sheet are conducted by the pressure roller.

Kutsuna et al. produced Laser Roll Welidng equipment experimentally, and conducted basic studies of the welding of low carbon steel to aluminum alloy (Rathod &Kutsuna, 2003, 2004). As a result, when the IMC layer thickness was from 4 to 8 μm, failure of base metal sheet was obtained after tensile shear test of specimen.

Until now, it's found that Laser Roll Welding has two types of welding mechanisms. The combination of iron and aluminum is that of metals with large difference in their melting points (Ozaki & Kutsuna, 2007). The combination of titanium and aluminum also corresponds to this category (Ozaki et al., 2008). In this case, only the metal with a higher melting point is heated up to an elevated temperature for avoid the formation of brittle IMC's. For example, in the case of steel to aluminum joint, steel is heated up to 1200 °C, because the max formation temperature of Fe_2Al_5 is 1169 °C. The interlayer is formed and cooled rapidly to minimize the thickness of brittle IMC's.

The other is a combination of metals with eutectic reaction at a lower temperature than melting point. For instance, the combination of titanium and iron falls into this type (Ozaki et al., 2007). In this case, the eutectic reaction occurs at a lower temperature, about 1085 °C, than melting points of both metals. The interlayer of the eutectic phase, β-Ti and TiFe, is formed at the interface.

In this way, uncoated materials have been mainly used in the previous study of Laser Roll Welding. However, coated materials such as zinc coated steel have not investigated enough yet. In this chapter, Laser Roll Welding of zinc coated steel and 6000 series aluminum alloy was conducted, and the weldability was investigated. Two types of zinc coated steel were used. One is hot-dip galvanized steel called as GI, the other is hot-dip galvannealed steel called as GA. The former is used mainly in Europe, the latter is in Japan. Then, the influences of process parameters, such as welding speed and roll pressure, on the formation of intermetallic compound layer and the change of zinc coated layer have been investigated to get sound joints with these galvanized steel. Furthermore, the effects of process parameters on joint performance have been also discussed.

2. EXPERIMENTAL PROCEDURE

2.1. Materials Used

As materials, two types of zinc coated steel and A6000 series aluminum alloy, Al-0.5Mg-1.0Si, were used for joining. The zinc coated steels were hot-dip galvanized steel, hereafter, called as GI, and hot-dip galvannealed steel, hereafter, called as GA, respectively. The dimension of the zinc coated steel sheets were 125 x 180 x 0.55 mm and that of the A6000 series aluminum alloy sheet, hereafter, called as A6000, was 125 x 180 x 1 mm. The zinc coating weight of the GI sheet was 60 g/m^2 and the GA was 45 g/m^2.

The surface of the steel sheets was coated with graphite by using a graphite spray to increase the absorption rate of laser beam. The thickness of the coating layer was approximately 10 μm. The faying surface of the steel sheets was only degreased with ethanol alcohol. The faying surface of the A6000 sheet was wire-brushed, polished by emery paper #600 and cleaned by ethanol before welding. Then, to remove the oxide firm on aluminum alloy sheet, the surface of the aluminum alloy was spread with flux, $KAlF_4{:}K_2AlF_5 \cdot H_2O$, 17-25 wt%, with particle size of 15 to 21 μm.

The setup of specimens is shown in Fig.4. A zinc coated steel sheet is placed as a top plate; an A6000 sheet is placed as a bottom with 3 mm width overlapping.

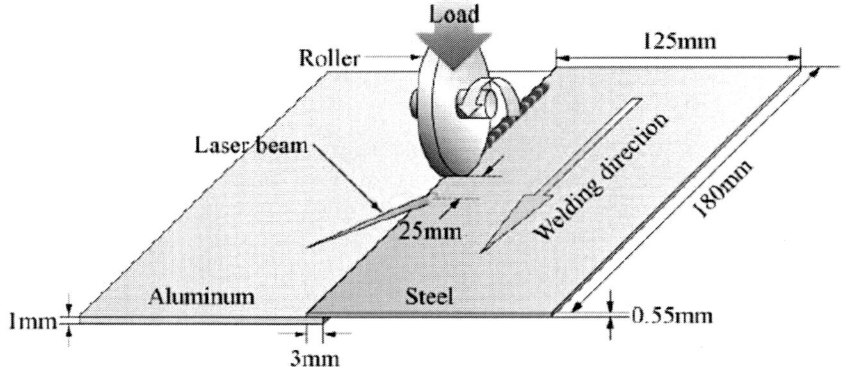

FIGURE 4. Setup of specimens in Laser Roll Welding

TABLE 2. Process parameters for Laser Roll Welding

Laser type	CO_2 laser (Pulse mode)
Laser peak power	2.0 kW
Duty cycles	75%
Frequency	150 Hz
Beam spot size	Minor axis : 2.5 mm Major axis : 3.5 mm
D_{LR}	25 mm
Welding speed	3.3 ~ 15.0 mm/s
Overlapped width	3 mm
Roll pressure	100 ~ 175 MPa
Shielding gas	Ar : 25 l/min

D_{LR}: Distance between Laser beam spot and Roller

2.2. Process Parameters

In this study, to investigate the effect of the welding condition on the joint properties of two types of zinc coated steel and aluminum alloy by Laser Roll Welding, welding speed and roll pressure were varied. Process parameters are shown in Table 2.

Pulsed laser was used by controlling a 2.4 kW CO_2 laser with continues wave. Laser peak power, duty cycles and frequency were constant 2.0 kW, 75 % and 150 Hz respectively. These parameters have been optimized in the previous study. Beam spot shape was an quasi-elliptical with major axis of 3.5 mm and minor axis of 2.5 mm. Distance between the center of laser beam spot and the pressurizing axis of roller was 25 mm. Welding speed was varied from 3.3 to 15.0 mm/s. Overlapped width was 3 mm and laser beam was irradiated the center of overlap. The roller is mounted with a calibrated compression spring for applying predetermined roll pressure. The roll pressure was calculated by assuming that the contact area between the roller and the steel sheet surface was a rectangle of 15 mm^2. Roll pressure was varied from 100 to 175 MPa. For the sake of the protection of condenser lens and the oxidation prevention of Laser Roll Welded joints, argon gas with flow rate of 25 l/min was used.

2.3. Evaluation Method Of Laser Roll Welded Joints

After welding, Laser Roll Welded specimens were cut across the lap joint seam for macrostructure and microstructure observation. Etching with 3% nital was made for observation of the interface layer. Hardness test and electron-probe microanalysis (EPMA) were conducted to analyze the interface layer and to identify the IMC present.

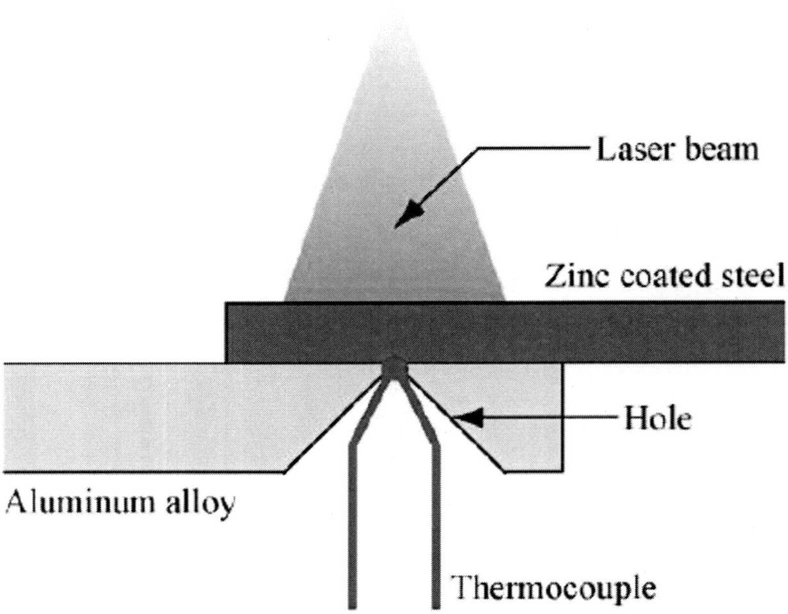

FIGURE 5. Schematic diagram of measurement method for thermal cycle at weld interface

Thermal cycle was measured using a Pt-PtRh thermocouple with a diameter of 0.3 mm. As shown inFig.5, a cone-shaped hole of 2 mm in diameter was drilled at the center of the overlap width in the aluminum sheet. The hole was used for placing thermocouple at the lower surface of the steel sheet. In addition, tensile shear test and Erichsen cupping test were carried out to investigate the mechanical properties of welded joint. The specimens with

20 mm width were used for tensile shear test and with 77 mm square for Erichsen cupping test.

3. EXPERIMENTAL RESULTS AND DISCUSSIONS

3.1. Bead Appearance And Cross-Section Of Laser Roll Welded Joints

Bead appearance of Laser Roll Welded specimen with the welding speed of 8.3 mm/s and roll pressure of 150 MPa are shown in Figs.6-7. Top bead is shown in Fig.6, and bottom bead is in Fig.7. Furthermore, cross-section of the weld bead is shown in Fig.8. In these figures, (a) shows GI/A6000 joint and (b) shows GA/A6000, respectively.

As shown in Fig.6, both of the bead turned black due to the effect of the graphite spray and the laser heating. In Fig.6 (a), striped pattern is observed on the GI sheet near the weld bead. This is the first time this striped pattern has been observed in the past study of Laser Roll Welding, the pattern seems to be unique to this welding of galvanized steel sheet. A GI sheet has galvanized layer, which exists as a thin layer at the surface of steel sheet. Thus it's thought that the pattern was caused by the laser heating and the roller pressure. On the other hand, the striped pattern doesn't exist on the GA sheet as shown in Fig.6 (b). This is because the coated layer of a GA sheet is alloyed zinc with iron.

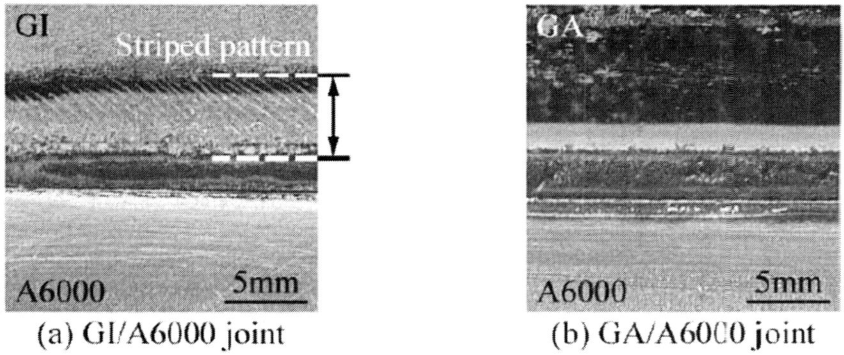

FIGURE 6. Top bead appearance of GI/A6000 and GA/A6000 Laser Roll Welded joints

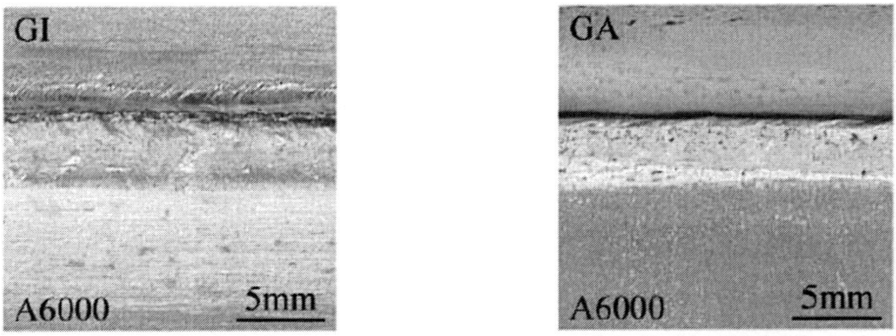

(a) GI/A6000 joint (b) GA/A6000 joint

FIGURE 7. Bottom bead appearance of GI/A6000 and GA/A6000 Laser Roll Welded joints

(a) GI/A6000 joint (b) GA/A6000 joint

FIGURE 8. Macro cross-section of GI/A6000 and GA/A6000 Laser Roll Welded joints

As shown in Figs.7-8 (a), partial melting and spreading of molten aluminum alloy on the GI sheet occurs in the A6000 sheet. The GI sheet was heated by laser, and the heat transferred to A6000 sheet by the contact of GI and A6000 sheets. Hence A6000 sheet was supposed to be melted. In contrast, molten A6000 alloy doesn't spread on the GA sheet.

In order to compare the wettability of the GI sheet with that of the GA sheet, bonding width was measured from the cross-section of the weld bead as shown in Fig.9 (a). Effect of the welding speed and the roll pressure on the bonding width of GI/A6000 and GA/A6000 joints are shown in Fig.9 (b) and (c), respectively.

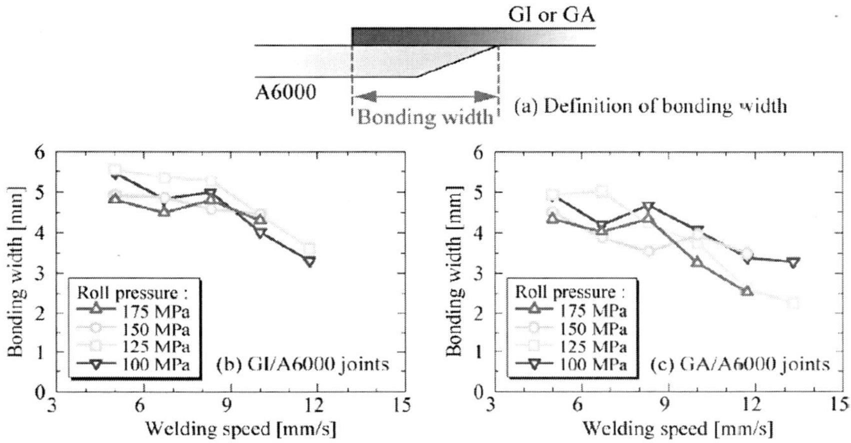

FIGURE 9. Effect of welding speed on bonding width at different roll pressures

It's common with Fig.9 (b) and (c) that the bonding width decreased as the welding speed is increased from 5.0 to 11.7 mm/s. This is because heat input decreases as the welding speed increased; the meltage of A6000 sheet also decreased. On the whole, the bonding width of GA/A6000 joints is narrower than that of GI/A6000. For example, the width with the welding speed of 8.3 mm/s and roll pressure of 150 MPa of GI/A6000 joint is 4.6 mm, and that of GA/A6000 is 3.5 mm. This is considered to be produced from the difference in the wettability of the surface of the GI and GA sheet.

3.2. Microstructures at Weld Interface

Interface microstructures of GI/A6000 joint with welding speed of 6.7 mm/s and roll pressure of 150 MPa are shown in Fig.10.

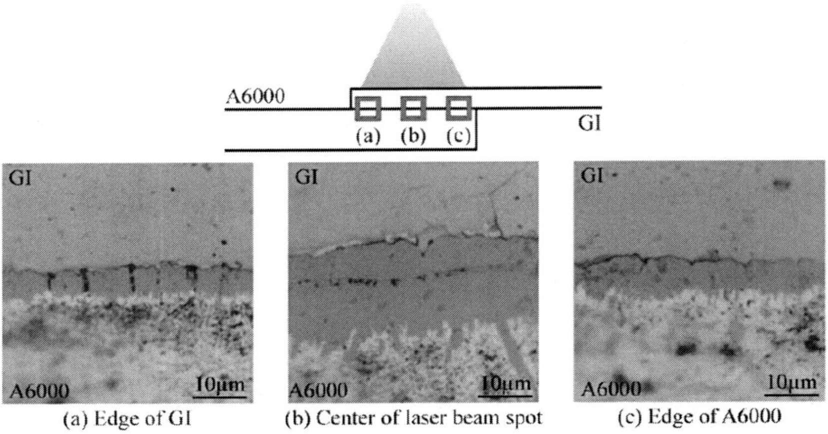

(a) Edge of GI (b) Center of laser beam spot (c) Edge of A6000

FIGURE 10. Microstructures at joint interface of GI and A6000 sheet

IMC layer was confirmed at the joint interface between the GI and the A6000 sheet. The IMC layer at the edge of the GI sheet is shown in Fig.10 (a), at the center of the laser beam spot is shown in (b) and at the edge of the A6000 sheet is show in (c). The IMC layer thicknesses in each place were 4.1 μm, 13.0 μm, and 4.9 μm respectively. Therefore, the thickness right under the center of laser beam spot was the thickest. Due to the high energy density near the center of the beam spot of the CO_2 laser used, it's thought that the area under the center of the beam spot is heated the highest temperature and cooled the slowest speed.

Interface microstructures of GA/A6000 joint with welding speed of 8.3 mm/s and roll pressure of 150 MPa are shown in Fig.11. IMC layer was also observed at the interface between the GA and the A6000 sheet. In addition, the thickness right under the center of laser beam spot was the thickest, and this was the same as that of the case of GI/A6000 joint.

3.3. Electron-Probe Microanalysis (EPMA) Of Interlayer

As the optical microscope doesn't reveal the details of the IMC layers, EPMA of iron, aluminum and zinc across joint interface were made to identify IMC's and existence of zinc. The results for GI/A6000 and GA/A6000 joints at different welding speeds with roll pressure of 150 MPa are shown in Figs.12-13. Bottom pictures show the SEM images of the IMC layer. In these figures, the EPMA results with the welding speed of 8.3 mm/s are shown in (a), with the welding speed of 10.0 mm/s are shown in (b).

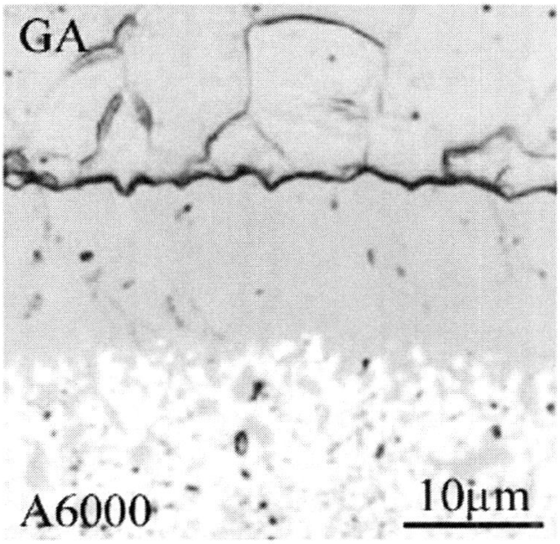

FIGURE 11. Microstructure at joint interface of GA and A6000 sheet

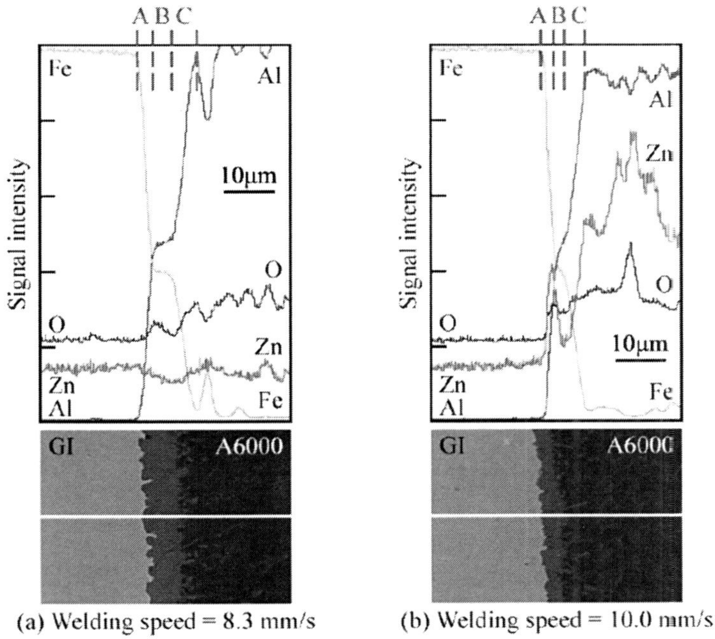

FIGURE 12. Results of EPMA at interlayer of GI/A6000 joints

From EPMA results of iron, aluminum and zinc across interface, IMC's and existence of zinc were identified. At the center of these EPMA results which are shown by the layer A to C in Fig.12 as an example, the signal intensity of Fe and Al has changed similarly. At the layer A, the intensity of Fe decreases rapidly, and Al rises. At the layer B, the IMC's are observed as stepped lines. At the layer C, the intensity of Fe decreases further and Al rises. From the estimation of composition from stepped lines at the layer B, it's suggested that main IMC's were brittle $FeAl_3$ and Fe_2Al_5.

When the welding speed is increased to 10.0 mm/s, zinc can be seen in aluminum alloy. From this result, it's thought that zinc tends to diffuse into aluminum when the welding speed becomes more than 10.0 mm/s. This reason is considered as follows.

Most of the zinc layer is vapored by the laser heating from the faying surface of the GI sheet at slow welding speeds. On the other hand, at fast welding speeds, the diffusion amount of the zinc into aluminum alloy by the roll pressure is larger than the evaporation amount of the zinc by the laser heating.

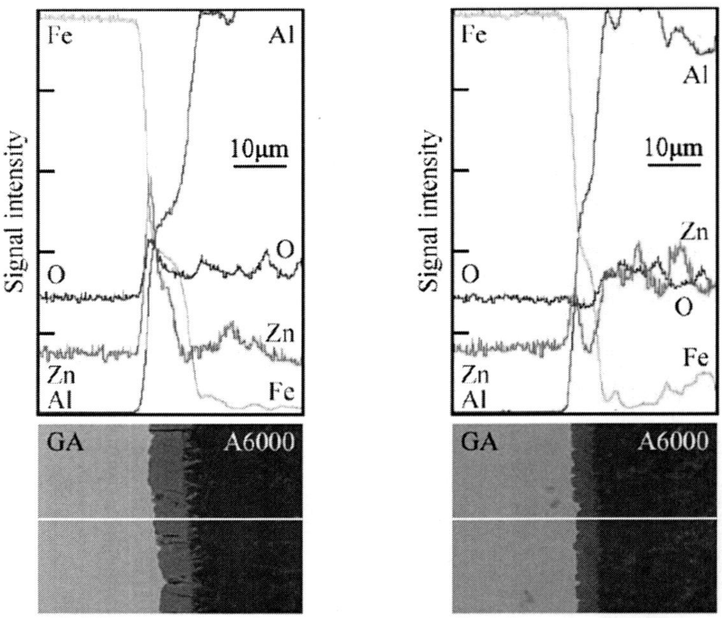

(a) Welding speed = 8.3 mm/s (b) Welding speed = 10.0 mm/s

FIGURE 13. Results of EPMA at interlayer of GA/A6000 joints

As shown in Fig.13 (a), the zinc line has higher peak than any other zinc line at the interface of the GA sheet side. This is because the zinc on GA sheet surface exists as Fe-Zn alloy. Then it's suggested that zinc was hard to be evaporated by the laser heating, and a lot of zinc was remained at the interface of the GA sheet side.

3.4. Vickers Hardness Measurements

From the EPMA results, it was presumed that main IMC's were brittle $FeAl_3$ and Fe_2Al_5. However, in order to obtain further evidence, Vickers hardness measurement was conducted. SEM image of GI/A6000 and GA/A6000 weld interface after the measurement with welding speed of 8.3 mm/s and roll pressure of 150 MPa are shown in Fig.14 (a) and (b), respectively.

Indentation size became small in order of the base metal of A6000, zinc coated steel, and the IMC layer. In particular, the indentation size of the IMC layer is much smaller than that of both base metals. As shown in Fig.14 (a), the Vickers hardness of the base metal of the GI, the A6000, and the IMC layer were 137 Hv, 94 Hv and 940 Hv on average respectively. The results show a large difference between IMC layer hardness and the surrounding base metals. The IMC hardness was about 10 times more than the A6000 base metal and 7 times more than the GI base metal. In the same way, as shown in Fig.14 (b), the Vickers hardness of the base metal of the GA, the A6000, and the IMC layer were 141 Hv, 97 Hv and 857 Hv on average respectively, the IMC layer was the hardest. Because the hardness of these IMC layers are between 892 Hv of $FeAl_3$ and 1013 Hv of Fe_2Al_5 according to Table 1, it's thought that $FeAl_3$ and Fe_2Al_5 are mainly formed at the interface as above.

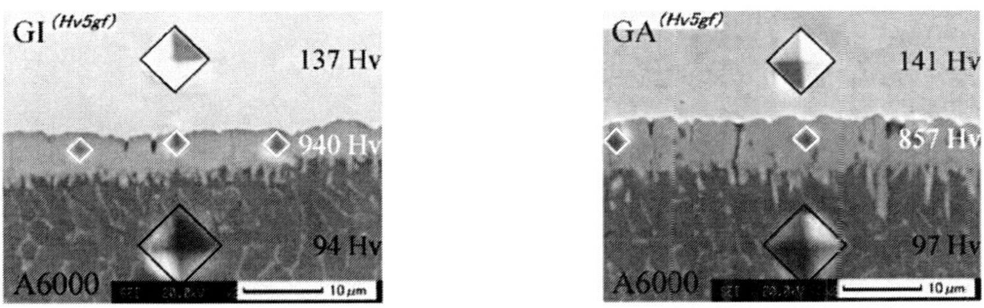

(a) GI/A6000 joint (b) GA/A6000 joint

FIGURE 14. Results of Vickers hardness measurement at joint interface

3.5. Effect of welding speed on thickness of intermetallic compound layer

Effect of the welding speed on the IMC layer thickness of GI/A6000 and GA/A6000 joints at different roll pressure are shown in Fig.15 (a) and (b), respectively. The thicknesses were measured right under the center of the laser beam spot.

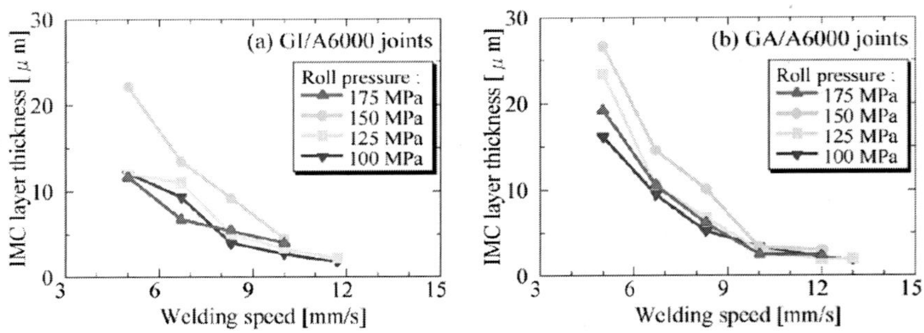

FIGURE 15. Effect of welding speed on IMC layer thickness of Laser Roll Welded joints interface

The IMC layer thickness decreases significantly as the welding speed is increased from 5.0 to 11.7 mm/s regardless of the roll pressure as shown in Fig.15 (a) and (b). This is because heat input decreases as the welding speed in increased. Therefore, this result indicates that the IMC layer thickness could be suppressed by heat input as was mentioned in previous study of Laser Roll Welding of steel to aluminum alloy.

3.6. Effect of roll pressure on thickness of intermetallic compound layer

Effect of the roll pressure on the IMC layer thickness of GI/A6000 and GA/A6000 joints at constant welding speed of 8.3 mm/s is shown in Fig.16.

The IMC layer thickness of both joints increases as the roll pressure is increased from 100 to 150 MPa. This is because increment of the roll pressure augments the contact between the zinc coated steel and A6000 sheet in this region. In contrast, the IMC layer thickness decreases as the roll pressure is increased from 150 to 175 MPa. This reason is considered as follows.

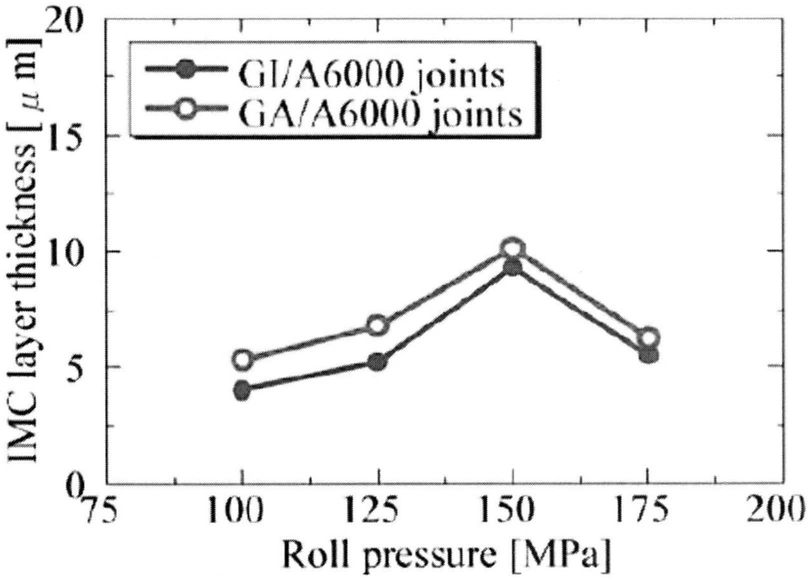

FIGURE 16. Effect of roll pressure on IMC layer thickness of Laser Roll Welded joints interface

The zinc coated steel and the A6000 sheet contact most widely at roll pressure of 150 MPa. However, when the roll pressure becomes more than 150 MPa, the steel sheet might be curved greatly by high roll pressure. Thus, it's thought that the contact area between the zinc coated steel and the A6000 sheet is narrowed in this region.

3.7. Effect of types of zinc coated steel on thickness of intermetallic compound layer

As shown in Fig.16, the IMC layer thickness of GA/A6000 joints is thicker than that of GI/A6000. This difference is attributed to the evaporation of zinc on the surface of each zinc coated steel sheets.

The boiling point of aluminum, iron and zinc are 2477 °C, 2887 °C and 906 °C respectively, that of zinc is far below those of aluminum and iron. Hence, the zinc on the surface of zinc coated steel sheet is melting, heat of fusion of 7.12 kJ/mol, and evaporating, heat of vaporization of 113.4 kJ/mol (The Japan Institute of Metals, 2004), in the process of Laser Roll Welding. However, because of the zinc on the surface of the GA sheet exists as Fe-Zn alloy, it's hard to be evaporated by the laser heating. Thus, laser energy is little-used for the zinc evaporation, and heat is conducted to A6000 sheet. On the other hand, the zinc on the surface of the GI sheet is easier to be evaporated than that of the GA. Therefore, it's thought that more laser energy for the zinc evaporation is used, and the heat conduction to A6000 sheet decreases.

3.8. Discussion of formation of intermetallic compound layer by thermal cycle at weld interface

Thermal cycle at the weld interface was measured to discuss about the effects of the welding speed and the roll pressure on the IMC layer thickness of Laser Roll Welded joints. The results of GI/A6000 and GA/A6000 joints at constant roll pressure of 150 MPa is shown in Fig.17 (a) and (b), respectively.

When the welding speed increases from 5.0 to 11.7 mm/s, different thermal cycles were obtained as shown in Fig.17 (a) and (b). Then, in order to quantitative the shape of these thermal cycles, peak temperature and holding time more than 500 °C were focused on.

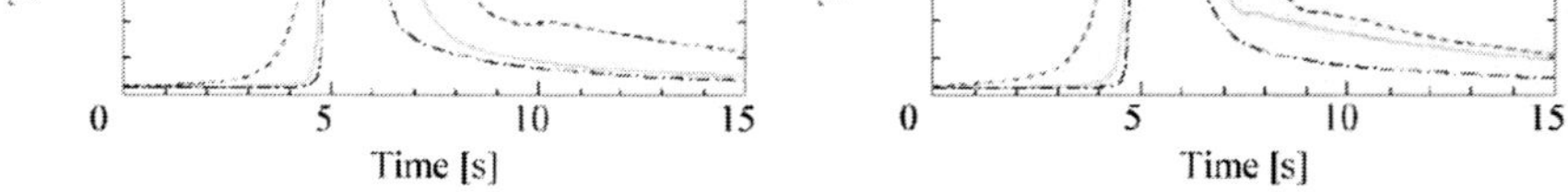

FIGURE 17. Effect of welding speed on interface thermal cycles of Laser Roll Welded joints

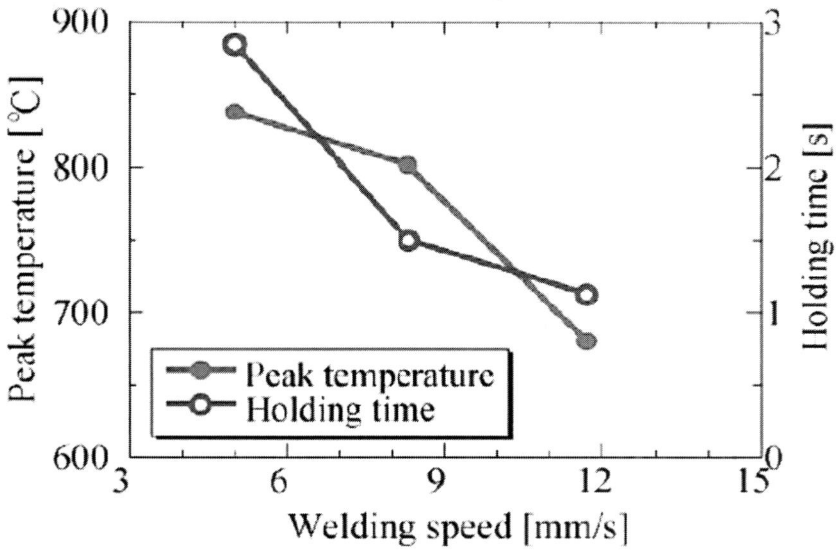

FIGURE 18. Effect of welding speed on peak temperature and holding time more than 500 °C at interface of GI/A6000 joints

The peak temperature and the holding time more than 500 °C of the thermal cycles in Laser Roll Welding of the GI and the A6000 at different welding speeds are shown in Fig.18. When the welding speed is increased from 5.0 to 11.7 mm/s, the peak temperature decreases from 850 to 680 °C, and the holding time more than 500 °C shortens from 2.8 to 1.1 sec at the weld interface. Hence the reduction of the IMC layer thickness by the increment of welding speed was attributed to the decline of the peak temperature and the shortening of the holding time. The similar results were obtained in Laser Roll Welding of the GA and the A6000.

When the welding speed is slow, there is excessive heat input and the cooling rate is slow. This provides surplus time for the formation of a thick interlayer containing a large amount of the Al-rich brittle IMC's When the welding speed is fast, there is suitable heat and time for melting of the aluminum and the diffusion process to take place. Therefore, it's thought that the change of the thermal cycle at the interface affect the formation of the IMC layer when the welding speed is varied.

In addition, when Fig.17 (a) is compared with (b), the peak temperatures of GA/A6000 joints are higher than those of GI/A6000. This fact caused the formation of thicker IMC layer in GA/A6000 joints than in GI/A6000 as shown in Fig.16.

Effect of the roll pressure on the interface thermal cycles of GA/A6000 joints at constant welding speed of 8.3 mm/s is shown in Fig.19. Thermal cycles are shown in Fig.19 (a) and the peak temperature and the holding time more than 500 °C of them are in (b). The reason of the changing of IMC layer thickness by the increment of the roll pressure as shown in Fig.16 is considered by using Fig.19.

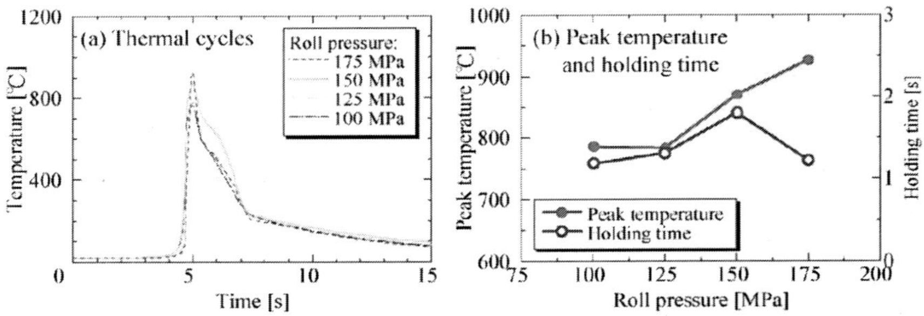

FIGURE 19. Effect of roll pressure on interface thermal cycles of GA/A6000 joints

From 100 to 150 MPa, the peak temperature and the holding time increase as the roll pressure is increased in Fig.19. Hence the IMC layer thickness increased as the roll pressure is increased inFig.16. This is because increment of the roll pressure augments the contact between zinc coated steel and A6000 sheet in this region as above. In contrast, from 150 to 175 MPa, the peak temperature rises, but holding time shortens as the roll pressure is increased in Fig.19. It's supposed that this tendency caused the decrement of IMC layer thickness as the roll pressure is increased at high region in Fig.16.

3.9. Results of tensile shear test

Tensile shear test was conducted to investigate the influence of the welding conditions on the weldability. The tensile shear specimens were prepared by cutting the welded specimen with 20 mm width.

Fig.20 shows the tensile shear specimen after testing of GI/A6000 joints. Failure in base metal of the GI sheet is shown in Fig.20 (a), and failure in interface is shown in (b). It's found that the base-metal-failure specimen failed far from the weld bead as shown in Fig.20 (a).

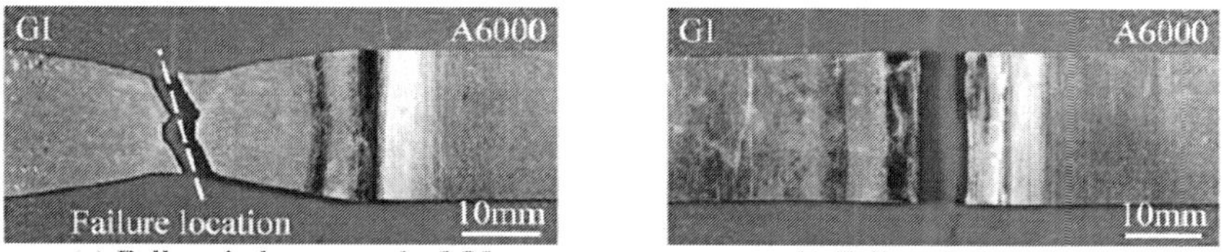

(a) Failure in base metal of GI (b) Failure in interface

FIGURE 20. Tensile shear specimen after testing of GI/A6000 Laser Roll Welded joints

Results of tensile shear test of GI/A6000 and GA/A6000 joints at various welding speed with the roll pressure of 150 MPa is shown in Fig.21 (a) and (b), respectively. Here the tensile shear strength was converted into the tensile shear load per millimeter of weld length, N/mm, as adopted by Peyre et al. (2007) and Sasabe et al. (2007). Failure in the base metal of the zinc coated steel sheet is shown as a circle in the figure, and failure in interface is shown as a solid mark.

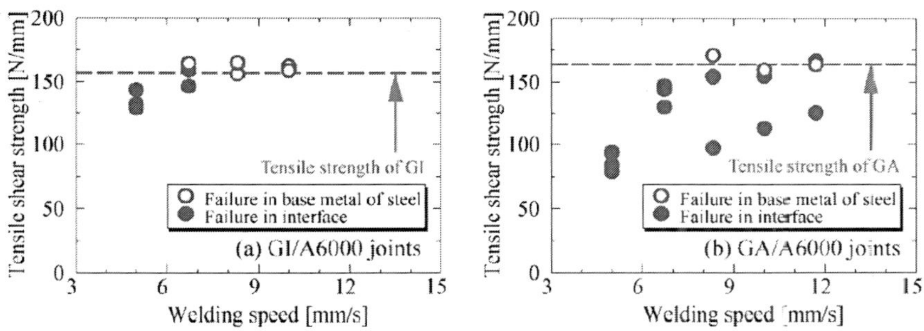

FIGURE 21. Effect of welding speed on tensile shear strength for Laser Roll Welded joints

When the welding speed is between 6.7 and 10.0 mm/s, the tensile shear strength shows high, and there are many specimens failed in the base metal. This result indicates that the IMC layer thickness is appropriate at these welding speeds. As shown in Fig.21 (a), specimens failed in the base metal could be obtained from 6.7 to 10.0 mm/s. In this region, the IMC layer thickness was less than 10 μm, and failure of specimen occurred in the base metal of the steel sheet. Therefore, it's thought that the IMC layer thickness should be less than 10 μm to get a good joint. As shown in Fig.21 (b), specimens failed in base metal were confirmed when the IMC layer thickness was less than 10 μm. This result corresponded to the articles by other researchers (Bruckner, 2005; Furukawa, 2005).

Additionally, when Fig.21 (a) is compared with (b), the tensile shear strength of GA/A6000 joints is lower than those of GI/A6000. There are three points which can be considered as this reason.

The first is the bonding width of GA/A6000 joints was narrower than that of GI/A6000 as shown inFig.9. Therefore, when tensile shear load was applied to the joints with narrow bonding width, the load concentrated to the weld. The second is the IMC layer thickness of GA/A6000 joints was thicker than that of GI/A6000 as shown in Fig.16. Moreover, the IMC layers were mainly composed of Al-rich brittle IMC's from the results of the EPMA and the Vickers hardness. The third is the residual zinc at the interface of GA/A6000 joints influenced the tensile shear strength. From the results of the EPMA, a lot of zinc was remained as Fe-Zn alloy layers at the interface of GA sheet side. The zinc coated layer of the GA sheet is formed from the steel side in order of Γ-phase (Fe_3Zn_{10}), Γ_1- (Fe_5Zn_{21}), δ- ($FeZn_7$) and ζ- ($FeZn_{13}$) (The Iron and Steel Institute of Japan, 1982). Since each phase has a difference in generation speed, the Γ-phase is thin and alloy layers are composed mostly of δ- and ζ-. Their phases are hard and brittle. Therefore, the interface of the GA/A6000 joints is weaker than that of the GI/A6000.

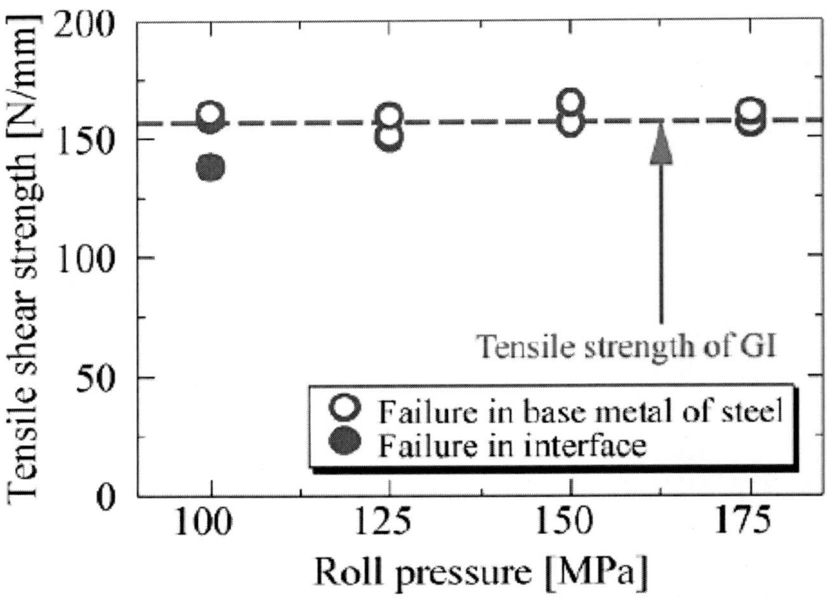

FIGURE 22. Effect of roll pressure on tensile shear strength for GI/A6000 joints

The maximum tensile shear strength of GI/A6000 joint, 162 N/mm, was obtained at the welding speed of 8.3 mm/s and the roll pressure of 150 MPa. This strength is equal to the tensile strength of the GI sheet, 157 N/mm, and 63% of the A6000 sheet, 256 N/mm. On the other hand, the maximum strength of GA/A6000 joint, 160 N/mm, was obtained at the welding speed of 10.0 mm/s and the roll pressure of 100 MPa. This strength is equal to the tensile strength of the GA sheet, 164 N/mm, and 62% of the A6000 sheet, 256 N/mm.

In addition, results of tensile shear test at various roll pressure with welding speed of 8.3 mm/s is shown in Fig.22. All specimens were failure in the base metal of the GI sheet except for the roll pressure 100 MPa. It's considered that the IMC layer thickness is less than 10 μm at these roll pressures as shown in Fig.16. Therefore, it seems that there is little influence of the roll pressure on the tensile shear strength.

3.10. Discussion of relation between intermetallic compound layer thicknessand tensile shear strength

From the above results, it has became clear that there is a close relation to the IMC layer thickness to the tensile shear strength also in Laser Roll Welding of zinc coated steel and aluminum alloy. There, the relationship between the IMC thickness and the tensile strength was discussed about all joints in this experiment. Effect of the IMC layer thickness on the tensile shear strength of GI/A6000 and GA/A6000 joints with different welding speed and roll pressure is shown in Fig.23 (a) and (b),respectively.

As above, only when the IMC layer thickness was less than 10 μm, the base-metal-failure specimens were obtained. When the IMC layer thickness is from 4 to 6 μm, high tensile shear strength could be obtained When the IMC layer is thicker than 6 μm, the tensile strength is decline. This is because the increment of the brittle IMC's at the joint interface might lead to weaken of the welded joint, and the joint isn't able to resist a heavy load.

On the other hand, when the IMC layer is thinner than 4 μm, the tensile strength also declines. In this case, due to low heat input might lead to incomplete welding at the joint interface. As shown in Fig.9, the bonding width is decreased by the decrement of the heat input.

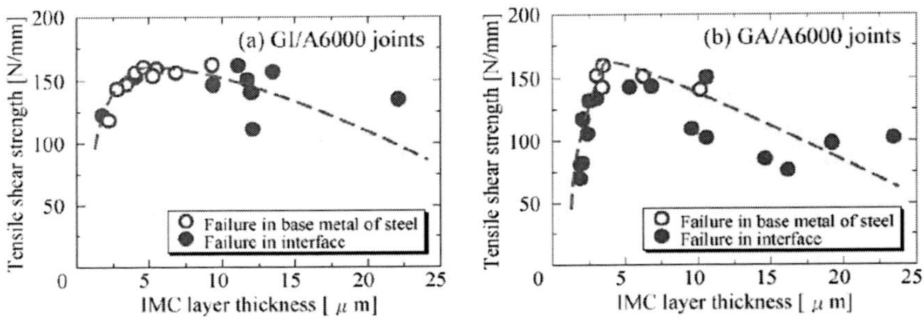

FIGURE 23. Effect of IMC layer thickness on tensile shear strength for Laser Roll Welded joints

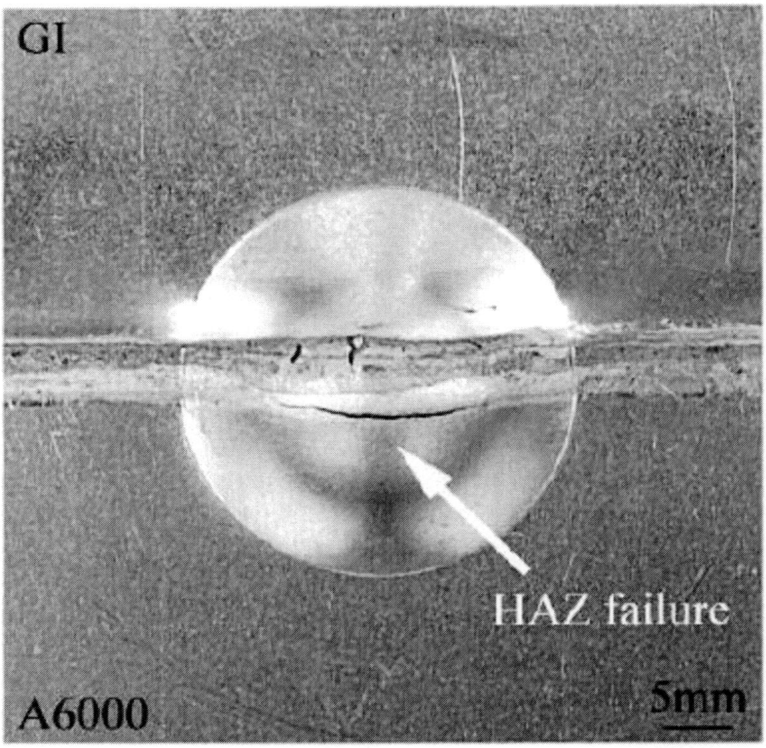

FIGURE 24. Erichsen cupping test specimen after testing of GI/A6000 joint

3.11. Results of erichsen cupping test

Finally, in order to investigate the formability of GI/A6000 and GA/A6000 joints, Erichsen cupping tests were carried out. The specimens were prepared by cutting the welded joints with the welding speed of 8.3 mm/s and roll pressure of 150 MPa into 77 mm square. A punch is pushed into the specimen. When the specimen is failed somewhere, the cupping height is evaluated as Erichsen value.

The Erichsen cupping test specimen after testing of GI/A6000 joint is shown in Fig.24. The specimen was failed at HAZ of the A6000 sheet side and the Erichsen value was 7.9 mm. With the GI base metal, the Erichsen value was 11.9 mm; with the A6000 base metal, the value was 8.6 mm. Therefore, this value was 92% of the base metal of A6000 sheet. In contrast, the specimen of GA/A6000 joint was failed at interface and the value was 3.6 mm. The same tendency was seen in the tensile shear test.

4. CONCLUSIONS

The present study is focused on the dissimilar metal joining of zinc coated steel and aluminum alloy by Laser Roll Welding. The following conclusions can be drawn.

1 The IMC layer was observed at the interface of all welded joins. It was suggested that most of the IMC's are brittle $FeAl_3$ and Fe_2Al_5 from the results of EPMA and Vickers hardness measurement. As the welding speed was faster than 10.0 mm/s, zinc was confirmed in aluminum alloy.

2 Increase in the welding speed led to decrease the bonding width and the IMC layer thickness at the joint interface. When the roll pressure was increased, the IMC thickness at the pressure of 150 MPa was the thickest. The IMC layer thickness of GA/A6000 joints was thicker than that of GI/A6000 on the whole.

3 Increase in the welding speed led to lowering of the peak temperature and shortening of the holding time more than 500 °C at the interface. The peak temperature at the roll pressure of 175 MPa was the highest, and the holding time at the pressure 150 MPa was the longest. The peak temperature of GA/A6000 joints was higher than that of GI/A6000 at the same welding condition.

4 When the IMC layer was less than 10 μm, failure of specimen occurred at the base metal of zinc coated steel in tensile shear test. The joint

properties of GI/A6000 joints were better than those of GA/A6000 from the results of tensile shear test and Erichsen cupping test.

5 The welding speed influenced the joint performance such as the IMC layer thickness and the tensile shear strength to a greater degree than the roll pressure in Laser Roll Welding.

ACKNOWLEDGEMENT

The authors would like to thank Mr. S. Nakagawa and Mr. K. Miyamoto from Research Center, Nissan Motor Co., Ltd. for their support in this research.

REFERENCES

1 J. Bruckner, 2005 Cold Metal Transfer Has a Future Joining Steel to Aluminum. Welding Journal, 84 6 3840 , 0043-2296

2 C. Dharmendra, K. P. Rao, J. Wilden, S. Reich, 2011 Study on Laser Welding-brazing of zinc Coated Steel to Aluminum Alloy with a Zinc Based Filler. Material Science and Engineering A, 528 14971503 , 0921-5093

3 J. Fan, C. Thomy, F. Vollertsen, 2011 Effect of Thermal Cycle of the Formation of Intermetallic Compounds in Laser Welding of Aluminum-Steel Overlap Joints. Physics Procedia, 12 134141 , 1875-3892

4 K. Furukawa, 2005 Welding Process of Iron-Aluminum. Welding Technology, 53 8 94102 , 0387-0197

5 S. Katayama, 2004 Laser Welding of Aluminum Alloys and Dissimilar Metals. Journal of Light Metal Welding & Construction, 42 1 1625 , 0368-5306

6 K. Katoh, H. Tokisue, 2004 Dissimilar Friction Welding of Aluminum Alloys to Other Materials. Journal of Light Metal Welding & Construction, 42 2 1118 , 0368-5306

7 T. B. Massalski, 1986 Binary Alloy Phase Diagrams 1 American Society for Metals, 148 0-87170-261-4

8 K. Nishimoto, K. Atagi, H. Fujii, Katayama, S. , 2005 Laser Pressure Welding of Dissimilar Metals Welding. Proceedings of the 63rd Laser Materials Processing Conference, 133138 , 4-94768-458-5 2005

9 O. Ohashi, 2004 Diffusion Bonding of Aluminum to Different Metals. Journal of Light Metal Welding & Construction, 42 2 1923 , 0368-5306

10 H. Okamura, K. Aota, 2004 Joining of Dissimilar Materials with Friction Stir Welding, Journal of Light Metal Welding & Construction, 42 2 110 , 0000-0368-5306

11 T. Okita, 2004 Resistance Welding of Aluminum Alloys to Dissimilar Metals. Journal of Light Metal Welding & Construction, 42 1 215 , 0368-5306

12 H. Ozaki, S. Hayashi, M. Kutsuna, 2008 Laser Roll Welding of Dissimilar Metal Joint of Titanium to Aluminum Alloy. Quarterly Journal of Japan Welding Society, 26 1 2430 , 0288-4771

13 H. Ozaki, R. Ichioka, M. Kutsuna, 2007 Laser Roll Welding of Dissimilar Metal Joint of Low Carbon Steel and Titanium. Quarterly Journal of Japan Welding Society, 25 1 173178 , 0288-4771

14 H. Ozaki, M. Kutsuna, 2007 Laser Roll Welding of Dissimilar Metal Joint of Low Carbon Steel to Aluminum Alloy Using 2kW Fiber Laser. Quarterly Journal of Japan Welding Society, 25 4 473479 , 0288-4771

15 P. Peyre, G. Sierra, F. Deschaux-Beaume, D. Stuart, G. Fras, 2007 Generation of Aluminum-steel Joints with Laser-induced Reactive Wetting. Material Science and Engineering A, 444 327338 , 0921-5093

16 M. Rathod, M. Kutsuna, 2003 Laser Roll Bonding of A5052 Aluminum Alloy and SPCC Steel. Quarterly Journal of the Japan Welding Society, 21 2 282294 , 0288-4771

17 M. J. Rathod, M. Kutsuna, 2004 Joining of Aluminum Alloy 5052 and Low-Carbon Steel by Laser Roll Welding. Welding Journal, 83 1 16s26s , 0000-0043-2296

18 S. Sasabe, T. Iwase, T. Matsumoto, Y. Hattori, T. Miono, 2007 Dissimilar Metal Joining of Aluminum Alloys to Steel by using Newly-developed Hot-dip Aluminized Steel Sheet. Journal of Light Metal Welding & Construction, 45 2 2333 , 0368-5306

19 D. Satou, 2004 Aluminum Clad of Explosive Welding. Journal of Light Metal Welding & Construction. 42 1 2630 , 0368-5306

20 The Iron and Steel Institute of Japan (1982). Iron and Steel Handbook Vol. 6 (3rd Edition), Maruzen, pp.421-434, Tokyo

21 The Japan Institute of Metals (2004). Metals Data Book (Revised 4th Edition), Maruzen, p.11, ISBN 4621073672, Tokyo

22 M. Torkamany, S. Tahamtan, J. Sabbaghzadeh, 2010 Dissimilar Welding of Carbon Steel to 5754 Aluminum Alloy by Nd:YAG Pulsed Laser. Materials and Design, 31 458465 , 0261-3069

23 S. Yamamoto, 2012 Welding in the World-Picking up from Foreign Magazine. Welding Technology, 60 3 (2012), 75 0387-0197

24 S. Yan, Z. Hong, T. Watanabe, T. Jingguo, 2010 CW/PW Dual-beam YAG Laser Welding of Steel/Aluminum Alloy Sheets. Optics and Lasers in Engineering, 48 732736 , 0143-8166

25 M. Yasuyama, K. Ogawa, T. Taka, 1996 Spot welding of aluminum and steel sheet with insert of aluminum clad steel sheet- Part 1. Quarterly Journal of Japan Welding Society, 14 2 314320 , 0288-4771

8

CHAPTER

In Situ Reaction During Pulsed Nd:YAG Laser Welding SiCp/A356 with Ti as Filler Metal

Kelvii Wei Guo

[1] MBE, City University of Hong Kong, Hong Kong,, China

1. INTRODUCTION

As aluminum matrix composites are generally low-cost and exhibit higher specific strength, high wearability, and good design performance property and functionality. They are widely applied in aerospace-flight, aviation structure, and automobile and in the heat resistant-wearable parts of engine [1-4]. Hence, a great deal of contemporary research effort is focusing upon their development and applications, typically on the discontinuously reinforced aluminum matrix composites like matrix with particle, short fiber, whisker and so forth. Additionally, a great deal of attention has also been drawn into the investigation of their secondary processing technologies like machining, joining and plastic forging. Welding is an important process for joining these materials. There has extensive effort to be devoted to developing appropriate process for joining the similar or dissimilar composites in literatures [2-4]. These processes can be mainly categorized as: (i) fusion welding like arc welding, shielding gas welding, laser welding and electron beam welding, *etc.*; and (ii) solid-state welding like soldering, explosive welding, friction welding and diffusion welding, *etc.* However, there still exist many problems in joining of the discontinuously reinforced aluminum matrix composites using conventional arc-welding processes and those high energy density welding methods like: laser welding and electron beam (EB) welding. These problems include: (i) the formation of poor weldment and the unsatisfactory properties of welded joints – mainly due to the high viscosity and poor flowability of the liquid welding pool causing mixing difficulty of the composite base material with filler materials; (ii) the occurrence of micro-segregation or inhomogeneous distribution of the reinforcement phases of SiC_p, Al_2O_{3p}, AlN and *etc.*, and whiskers like SiC_w – typically owing to the rejection by their solidification front in the welding poor as cooling down, which subsequently prompts for many micro and macro defects in the weld and very poor properties of the welded joints; (iii) the formation of aluminum carbide – mainly as a result of harmful interfacial reaction between aluminum matrix and reinforcement phases; and (iv) *so on.* A typical example of the interfacial reaction likely to have pernicious effects on the mechanical and chemical behavior of the composite is $4Al + 3SiC \rightarrow Al_4C_3 + 3Si$. This is owing to: (a) the formation of brittle and weak aluminum carbide Al4C3 in the interfacial reaction sacrifices the reinforcement-materials in the composite; and (b) the unstable aluminum carbide in wet environments causes corrosion of the composite because of its rapid hydrolysis *etc.*

Aiming at developing or improving the conventional welding technique, this paper studies the technique of welding the stir-cast aluminum matrix composite SiC_p/A356 by Nd:YAG laser welding with pure titanium as filler. This study has been specifically concerned on the *in situ* reinforcement effect of Ti on the microstructures of laser welded joints, which have been analyzed by means of Scanning Electron Microscope (SEM), Transmission Electron Microscope (TEM) and Electron Diffusion X-ray analysis (EDX) *etc.* The study aims at providing some ground works for further studies in this field.

2. EXPERIMENTAL MATERIAL AND PROCESS

2.1. Experimental material

Stir-cast SiC_p/A356 aluminum matrix composite, reinforced with 20 % volume fraction SiC particle of 12 μm mean size, was used as the welding specimens. The tensile strength of such specimens was 240 MPa and their solid-liquid phase transformation temperature was in the range of 562.6～578.3 ˚C.Figure 1 showed their corresponding microstructure while Table 1 tabulated the chemical composition of the A356 matrix alloy. Pure titanium was used as the filler metal.

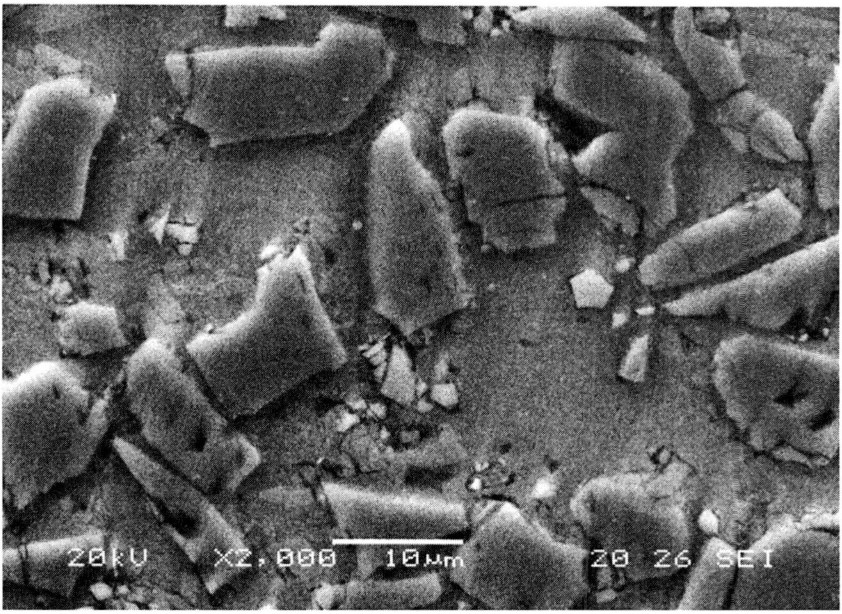

FIGURE 1. Microstructure of SiCp/A356 aluminum matrix composite

TABLE 1. Composition of A356

Composition (Wt %)			
Si	Mg	Ti	Al
6.5~7.5	0.3~0.5	0.08~0.2	Bal.

2.2. Experimental process

The stir-cast aluminum matrix composite specimens were individually wire-cut to the size of 3 mm × 10 mm × 35 mm (Fig. 2). The quench-hardened layer induced by wire-cut and the oxide on the surfaces of specimens were polished away by 400 # emery cloth. The pure titanium filler was then machined to depth 3 mm × width 10 mm × thickness of 0.15

mm, 0.3 mm, 0.45 mm, 0.5 mm, 0.6 mm and 0.75 mm, respectively The specimens were then mounted into a clamping devise on the platform of a GSI Lumonics Model JK702H Nd:YAG TEM_{00} mode laser system, and their welding surfaces were properly cleaned by acetone and pure ethyl alcohol so as to remove any possible contaminant. The prepared pure titanium filler was also thoroughly cleaned and carefully sandwiched between the two composite specimens in the clamp. Thereafter, specimens were welded immediately by the Nd:YAG laser with wavelength of 1.06 μm, defocused distance of 10 mm so as to give a focus spot diameter of approximately 1.26 mm, laser fluence energy 2 J, frequency 25 Hz, and pulse duration 4 ms. In the welding, the relative moving speed of the laser and the welding pieces (i.e. feedrate) was set at 300 mm/min

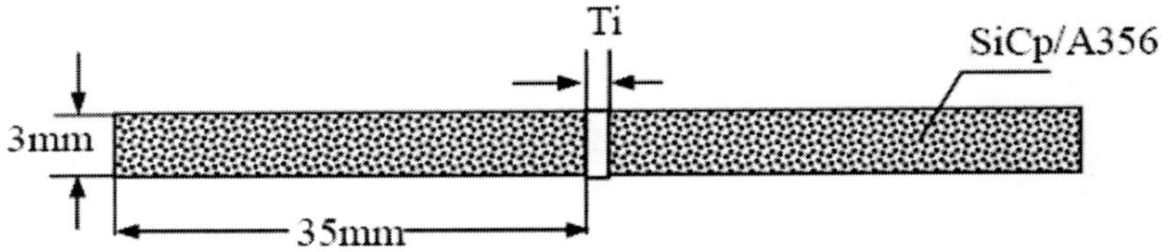

FIGURE 2. Schematic illustration of laser welding with Ti filler

Tensile strength of the joint was performed by a MTS Alliance RT/30 electron-mechanical material testing machine with a straining velocity of 0.5mm/min.The cross-section of welded joints was wire-cut for optical microscope investigation, and Scanning Electron Microscope (SEM) and Transmission Electron Microscope (TEM) analyses. SEM was used to analyze in detail the microstructure at the weld joints and the fractured tensile test-pieces of the joints. Optical microscope was used for observing the structure of a large area. TEM and Electron Diffusion X-ray analysis (EDX) were used to analyze the interface between the newly-formed phases and aluminum matrix, the distribution of chemical elements and spectra at the joints. The Nd:YAG laser with the similar setting conditions and feedrate was also used to weld the aluminum matrix composite specimens without any filler.

3. RESULTS AND DISCUSSION

3.1. Microstructures and properties of welded joints

The microstructure of the *in situ* reinforcement of Ti by Nd:YAG laser welding with 0.3 mm thick Ti filler was shown in Fig. 3. Appearance of *in situ* reinforcement particles distributed uniformly in the welded joint was seen. It also showed the disappearing of the drawbacks like incomplete fusion and pernicious phase Al4C3. These subsequently resulted in higher tensile strength (Table 2) of the joint. Comparatively, the reinforcement particles distributed more compactly than that of parent composite (cf. Fig. 1 and Fig. 3). The relatively more highly compacting reinforcement particles improved distinctly the properties of welded joints. The presence of Ti effectively improved the flowability of the liquid welding pool and the newly formed *in situ* reinforcement particles (Fig. 3) replaced those initial reinforcement particles (Fig. 1). Those dimples appeared in the SEM of the corresponding fractured surface (Fig. 4) suggested that: (i) the new-formed reinforcement particles had been perfectly wet; and (ii) the harmful composite structure of the initial welding viz. reinforcement/Ti/reinforcement had been changed to the state of reinforcement/matrix /reinforcement. XRD pattern of the fractured surface (Fig. 5) of the weld joint did not give sign of any harmful phase or brittle phase of Al4C3. This suggested the occurrence of effective interface transfers between reinforcement particles and matrix in the laser welded joint that subsequently provided favorable welding strength (Table 2). By the newly formed *in situ* reinforcement particles as detected by EDX (Table 3) and the intensity spectra shown in Fig. 5, the newly-formed reinforcement particle in the weld was identified as TiC.

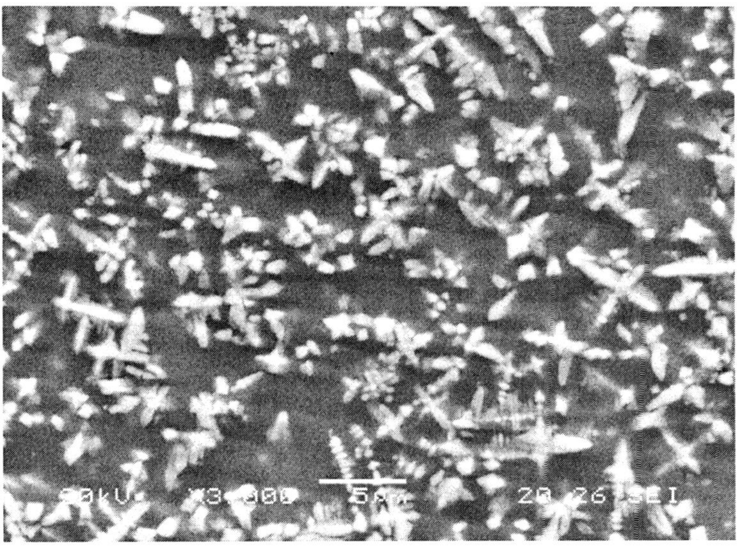

FIGURE 3. Microstructure of *in situ* reinforcement by laser welding with 0.3 mm thick Ti filler

TABLE 2. Mean tensile strength of laser welded specimens with various Ti filler thicknesses

Thickness of Ti filler (mm)	0.00	0.15	0.3	0.45	0.60	0.75
Mean strength (MPa)	76-91	126-135	168-180	143-160	107-113	79-96

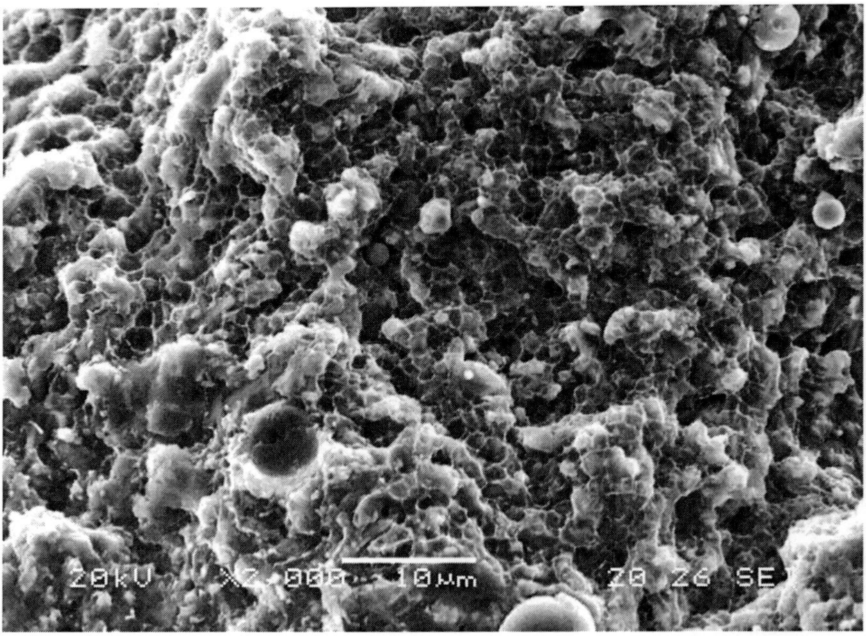

FIGURE 4. Fractograph of the laser welded joint with 0.3 mm thick Ti filler

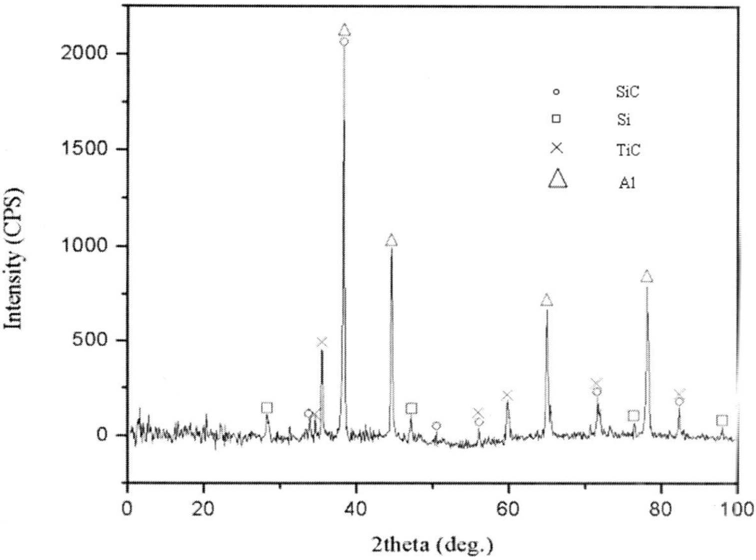

FIGURE 5. XRD pattern of the fracture surface for laser welding with 0.3 mm thick Ti filler

TABLE 3. EDX analysis of newly-formed particle in the laser weld with 0.3 mm thick Ti filler

Element	Ti	Si	Al
Wt (%)	81.84	4.42	Bal.

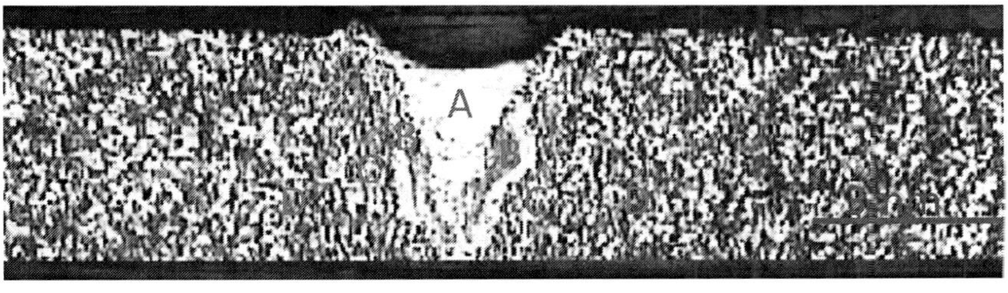

FIGURE 6. Macro-structure of the laser welded joint with 0.3 mm thick Ti filler

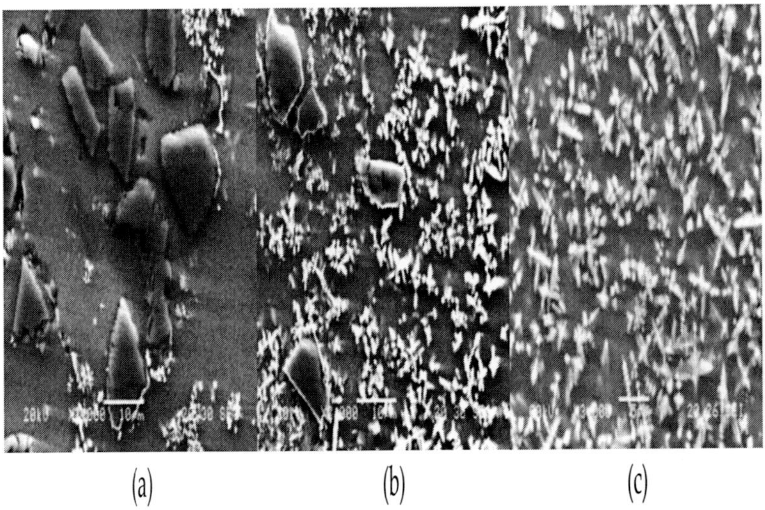

FIGURE 7. Microstructures of the different areas in the laser weld with 0.3 mm thick Ti filler(a) Area A (b) Area B (c) Area C

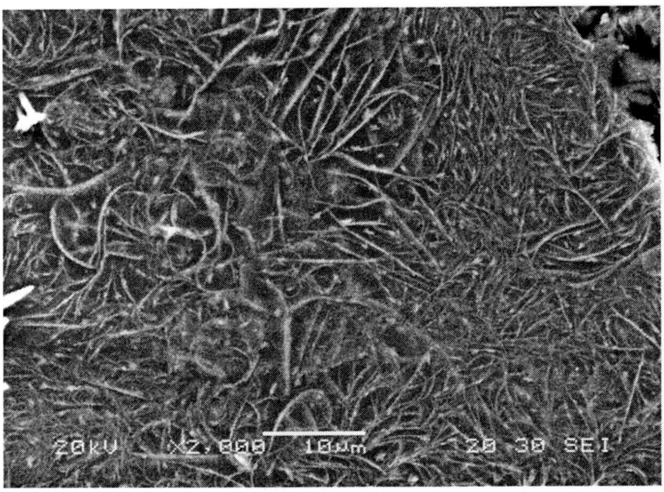

FIGURE 8. Microstructure of the weld with no-filler

Figure 6 illustrates the macro-structure of welded joint with Ti filler. Basically, the weld consisted of three main areas, namely: the *in situ* reinforcement area A, the two transitional areas B and C, and the no-*in situ* reinforcement area D. Their individual microstructures were shown in Fig. 7. The microstructures indicated the initial reinforcement SiC particles were completely replaced by the newly-formed *in situ* reinforcement TiC particles that mainly resulted in the formation of the *in situ*reaction area A (Fig. 7a). In the area B, the newly-formed *in situ* reinforcement TiC particles and the initial reinforcement SiC particles were coexistent (Fig. 7b). In the area C, very little newly-formed *in situ* reinforcement TiC particles were found (Fig. 7c). In the area D, there only existed the initial reinforcement particles SiC (Fig. 1). It was found that the typical pernicious acicular Al4C3 microstructure had been effectively alleviated in the welded area. Hence, it improved markedly the properties of the welded joints and their achievable tensile strength was up to 180 MPa (Table 2) that was about 75 % of the strength of SiCp/A356. The microstructure (Fig. 8) of the traditional Nd:YAG laser welding with no-filler showed that there were lots of pernicious acicular Al4C3 in the weld, which led to lower weld joint tensile strength (Table 2), i.e. only as high as 91 MPa (that was about 37.9 % of its parent SiCp/A356).

3.2. Element distribution in the transition area

Figure 9 illustrates the element distribution of the area B in the weld as shown in Fig. 6 and Fig. 7b. It showed the newly-formed *in situ* reinforcement particles were around the initial reinforcement SiC particles which offered a high density nucleus area for the nucleation of *in situ* TiC. The stirring effect in the welding pool by laser irradiation and the initial reinforcement SiC particles would be replaced by the newly-formed *in situ* TiC (cf. Figs 9b and 9c) following the chemical reaction process of: Ti (l) + SiC (s) —→ TiC (s) + Si (s)

According to the free energy for forming the SiC, TiC and Al_4C_3 carbides as elaborated in Fig. 10 of literatures [5, 6, 7], the free energy required to form TiC was much lower than that for Al_4C_3 when the reaction temperature was above 800 ºC. The affinity between Ti and C in the Nd:YAG laser welding was therefore greater than that of Al and C. The chemical reaction between Ti and SiC in the welding pool would subsequently take precedence over the reaction between Al and SiC and thus resulted in restraining the formation of the pernicious acicular Al_4C_3.

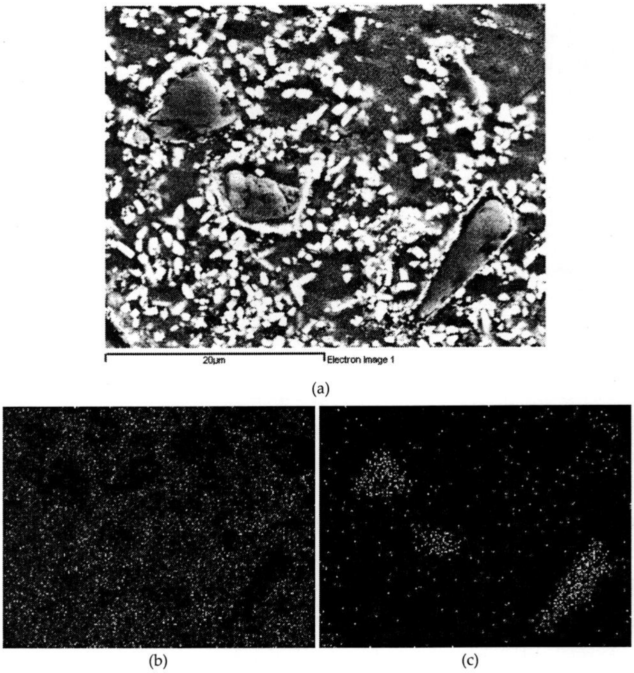

FIGURE 9. Element distribution of B area in the weld; (a) micrograph of the area B (b) Ti element face distribution (c) Si element surface distribution

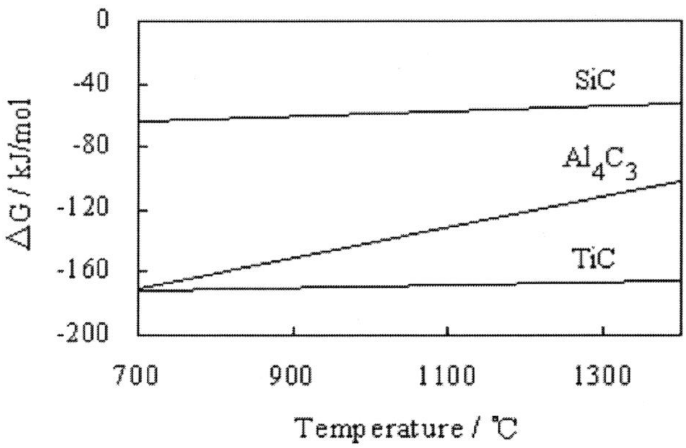

FIGURE 10. Free energy of formation of several metallic carbides [5, 6, 7]

3.3. Influence of ti filler thickness

The microstructures of *in situ* reinforcement with various thicknesses of Ti filler were illustrated inFig. 11 and its corresponding fractographs were shown in Fig. 12. It illustrated that the amount of the formed *in situ* TiC was distinctly increased with the increase in the thickness of Ti filler. Test indicated that maximum strength of welded joints (Table 2) was achieved at Ti filler thickness 0.3 mm (Figs. 3 and 4). This was because the newly-formed *in situ* reinforcement particles TiC were uniformly distributed in the weld and the initial irregular (mostly in hexagonal shape as shown in Fig. 1)reinforcement SiC particles in the weld were no longer observed (Fig. 3). Moreover, the pernicious acicular Al_4C_3 was successfully restrained (Figs 3 and 7a). At the thickness of Ti filler below 0.3 mm, there was little sign of the newly-formed *in situ* reinforcement TiC particles to be observed (Fig. 11a) and a number of pernicious acicular Al_4C_3 were formed in the weld. When the thickness of Ti filler was just beyond 0.3 mm, the properties of the joints tended to become poorer again (Fig. 12b). This was because the input energy was mainly used for melting the Ti filler and resulted in the coexistence of coarse columnar crystals and fine equiaxed crystals (Fig. 11b). When the thickness of Ti filler was further increased (Fig. 11c), the percentage of liquid Ti in the welding pool would also be increased. Subsequently, the weld zone would form coarser columnar crystals, as displayed in the SEM micrograph of Fig. 11c, after the resolidification of the melt. From the phase diagram (Fig. 13) of Ti-Al binary system [8], it can be anticipated that increasing the content of Ti would lead to the intermetallic compounds like TiAl and Ti_3A_l, *etc.*, to be formed during the Nd:YAG laser welding. As illustrated by the XRD pattern of the fractured surface of a laser weld joint with the thicker Ti filler inFig. 14, there were some distributing brittle intermetallic compounds like TiAl and Ti_3Al to be detected. Available literature [9] has demonstrated that TiAl and Ti_3Al are the harmful intermetallic compounds in the weld and tend to decrease obviously the strength of welded joints. Such harmful effect may follow the chemical reaction of: 5Ti[Al l] +3Al[l] + SiC[s]—→TiC[s] + Si[Al [l]] + Al[l] + (TiAl + Ti_3Al). Hence, too thick of the Ti filler led to: (i) the appearance of the large block of columnar crystals in the microstructure (Fig. 15); and (ii) the new-formed reinforcement TiC to be replaced by the melted/re-solidified Ti and subsequently only the melted/re-solidified Ti existing in the weld. Results (Figs 3, 7 and 11) indicated that there existed an optimal thickness of Ti filler in the individually set parameters in the Nd:YAG laser welding of SiCp/A356. With the optimal thickness of Ti filler, the initial reinforcement particles SiC distributed in the aluminum matrix composite SiCp/A356 would offer a highly dense nucleus area for the *in situ* TiC nucleation. This would effectively alleviate the forming of intermetallic compounds like TiAl and Ti_3Al in the weld. It was in this manner that ultimately created favorable condition to provide relatively superior strength of the welded joint to that of conventional laser welding.

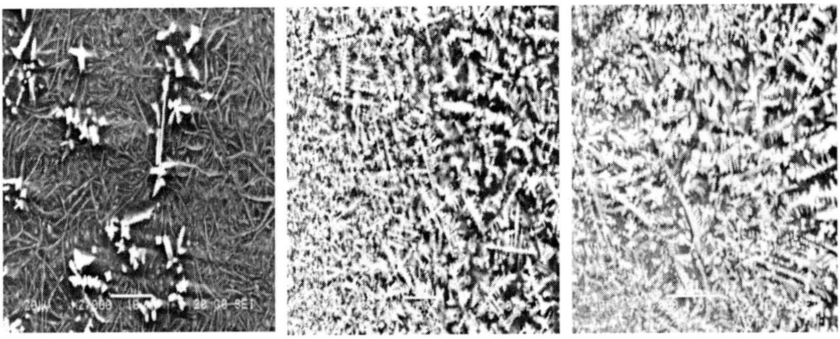

(a) (b) (c)

FIGURE 11. Microstructures in the A area with various thicknesses of Ti filler; (a) δ=0.15 mm (b) δ=0.45 mm (c)δ=0.60 mm

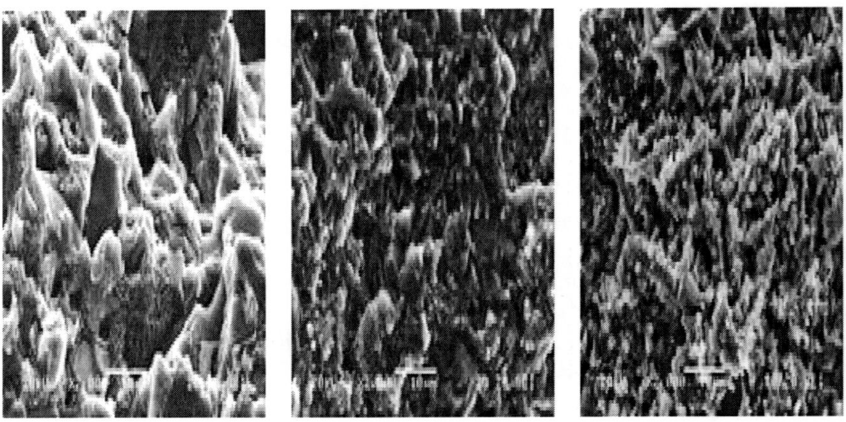

(a) (b) (c)

FIGURE 12. Fractographs of welded joints with various thicknesses of Ti filler; (a) δ=0.15 mm (b) δ=0.45 mm (c)δ=0.60 mm

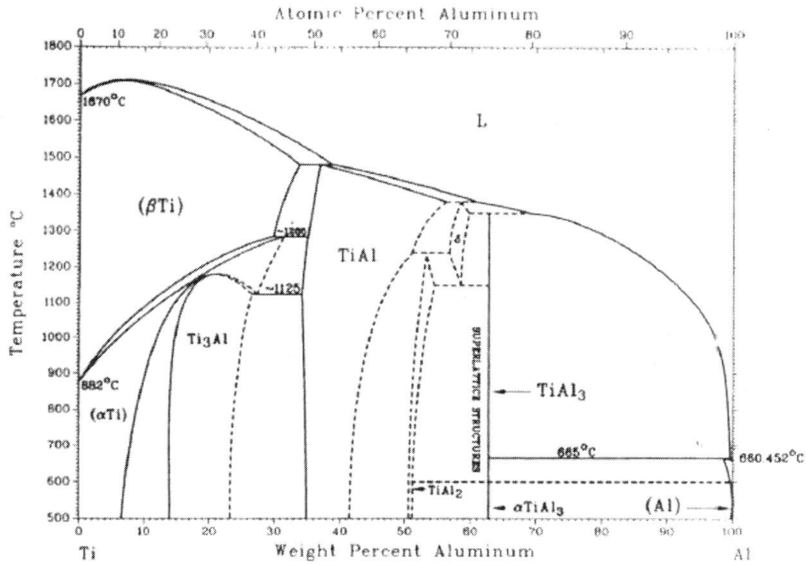

FIGURE 13. Binary phase diagram of Ti—Al [8]

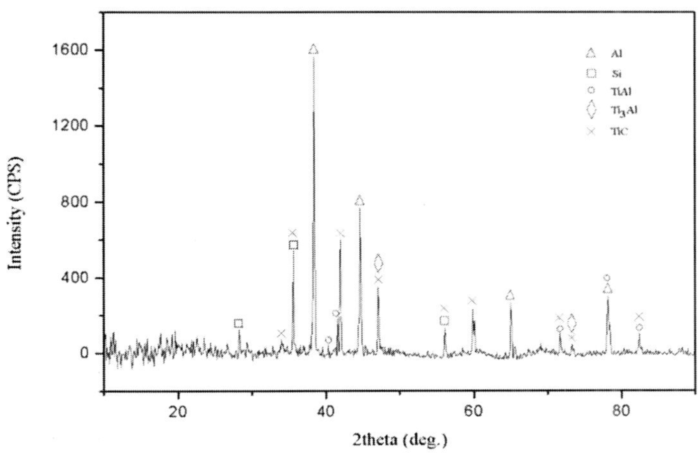

FIGURE 14. XRD pattern of fracture surface (δ=0.6 mm)

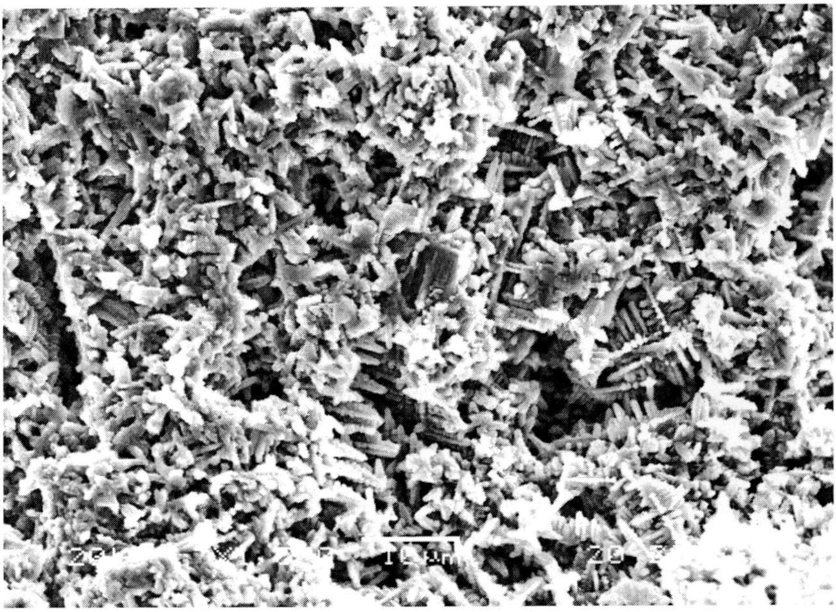

FIGURE 15. Columnar crystals in the laser weld with 0.6 mm thick Ti filler

3.4. Tem of the interface between *in situ* reinforcement particle tic and matrix

The interface between *in situ* reinforcement particle TiC and the matrix was analyzed by the TEM micrograph displayed in Fig. 16. It demonstrated a clear interface between the newly-formed *in situ*reinforcement particle TiC and matrix. This clearly distinct interface (Fig. 16) suggested the occurrence of prominent *in situ* reaction to integrate the reinforcement particle with matrix (cf. Figs 4and 16), and the high probability of successfully transferring load from the matrix to TiC and vice versa. It also gives indication that the aluminum matrix composite SiCp/A356 would be welded satisfactorily by Nd:YAG laser.

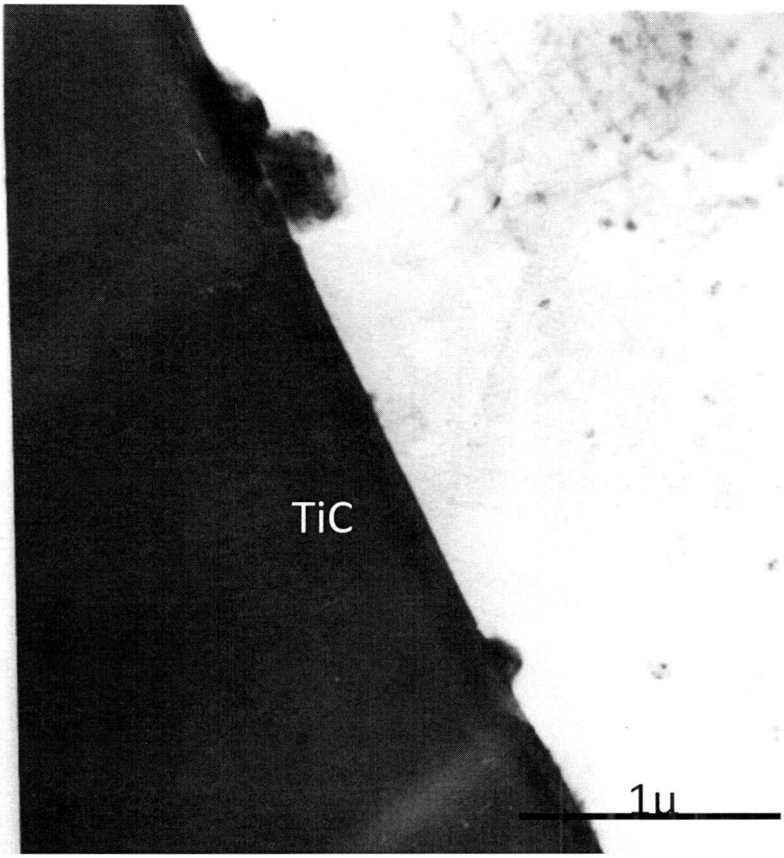

FIGURE 16. TEM of interface between *in situ* TiC reinforcement and the matrix for laser welding with 0.3 mm Ti filler

4. MICROSTRUCTURE EVOLUTION DURING THE WELDING

4.1. Temperature field of laser welding

The heat is assumed to be released instantaneously at time t=0 on the surface of the substrate. This causes a temperature rise in the material as follows [10-12]:

$$T - T_0 = \frac{Q}{\rho c(4\pi\alpha t)^{\frac{3}{2}}}\exp\left(\frac{-R^2}{4\alpha t}\right) \tag{1}$$

where ρ is the material density, C is specific heat, α is thermal diffusivity, λ is thermal conductivity, $Q$is the input energy.

When temperature distribution is quasi-steady state:

$$T - T_o = \frac{q_o}{2\pi\rho c\alpha R}\exp\left(-\frac{v}{2\alpha}(R + x)\right) \tag{2}$$

During the Nd:YAG laser welding, q_0 can be expressed as

$$q_0 = \eta(N(f, v)) \times \frac{P}{\frac{\pi D^2}{4}\times PD} = \eta(N(f, v)) \times \frac{4P}{\pi D^2 \times PD} \tag{3}$$

Where

η(N(f,v)) is the coefficient of laser welding input energy, which is direct proportional to number of overlaps or pulse frequency. With number of overlaps increasing, or pulse frequency increasing with constant the feed-rate, η(N(f,v)) will be increased synchronously. With the velocity (feedrate) increasing, number of overlaps with the constant pulse frequency will be decreased correspondingly led to lower heat input, η(N(f,v)) will be decreased accordingly.

Eq. (2) can be rewritten as:

$$T - T_o = \frac{2\times\eta(N(f,v))\times P}{\pi^2 D^2 \rho c\alpha R\times PD}\exp\left(-\frac{v}{2\alpha}(R + x)\right) \tag{4}$$

Defineξ=12πρcαξ=12πρcα, then Eq. (4) can be written as:

$$T - T_o = \frac{\xi \times \eta(N(f,v)) \times P}{\frac{\pi D^2}{4} \times PD \times R} \exp\left(-\frac{v}{2\alpha}(R+x)\right) \tag{5}$$

$$\text{Define} q = \frac{P}{\frac{\pi D^2}{4} \times PD}, \tag{6}$$

then Eq. (6) can be written as

$$T - T_o = \frac{\xi \times \eta(N(f,v)) \times q}{R} \exp\left(-\frac{v}{2\alpha}(R+x)\right) \tag{7}$$

4.2. Simulation model

4.2.1. EQUATIONS FOR TEMPERATURE DISTRIBUTION

Using energy balance, a differential equation can be obtained for the steady temperature distribution in a homogeneous isotopic medium, that is

$$\frac{\partial}{\partial x}\left(K_x \frac{\partial \theta}{\partial x}\right) + \frac{\partial}{\partial y}\left(K_y \frac{\partial \theta}{\partial y}\right) + \frac{\partial}{\partial z}\left(K_z \frac{\partial \theta}{\partial z}\right) = -q^B \tag{8}$$

Where the boundary conditions are $\theta|_{s_1} = \theta_e, K_s \frac{\partial \theta}{\partial x}|_{s_2} = q^s$.

For

$$\pi = \int_V \left\{ K_x \left(\frac{\partial \theta}{\partial x}\right)^2 + K_y \left(\frac{\partial \theta}{\partial y}\right)^2 + K_z \left(\frac{\partial \theta}{\partial z}\right)^2 \right\} d_V - \int_V \theta q^s d_V - \int_{s_2} \theta^s q^s \tag{9}$$

After Eq. 9 is discrete for the element, and according to $\delta\pi = \sum_{e=1}^{n} \delta\pi_e = 0$, it will be obtained

$$\overline{K^K}\bar{\theta} = \overline{\theta_s} + \overline{\theta_B} - \overline{C\theta} + \overline{K^c}(\theta_c - \overline{\theta^s}) + \overline{K^r}(\theta_r - \overline{\theta^s}) \quad (10)$$

where S: isothermal boundary, B: the heat-input, c: the conductive and r: the irradiative.

4.2.2. HYPOTHESIS AND MESH

Based on the situations during the laser welding and mainly focused on the temperature distribution, it is supposed that the laser resource is considered as a Gaussian distribution. Also, on the basis of specimen size wire-cut, the calculating size is set as 25 mm (x) × 20 mm (y) × 3 mm (z), the schematic of its finite element (FE) mesh is shown in Fig. 17. Moreover, Ti filler is considered as a section of the substrate with the different properties to ignore the effect of gap between the Ti filler and the substrate.

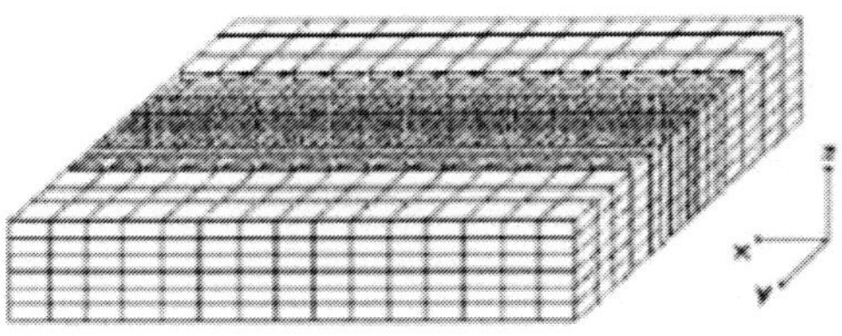

FIGURE 17. FE mesh for 3D numerical analysis

4.2.3. TEMPERATURE DISTRIBUTION

The simulated results are shown in Fig. 18 to Fig. 23. It shows that the temperature without Ti filler is same as the traditional laser welding. Simultaneously, due to the heat input into the substrates directly, without the additional heat resource for melting Ti filler, the peak of temperature (heat input) is relatively higher to form the weld. As a result, increasing the heat input into the substrate will decrease the tensile strength of the welded joint and wide the heat affected zone (HAZ) resulted in lower properties in the succedent practical applications (Table 2 and Fig. 19). Furthermore, a large amount of coarser acicular Al_4C_3 distributed in the fracture surface as shown in Fig. 19, which decreased the tensile strength of the welded joints seriously.

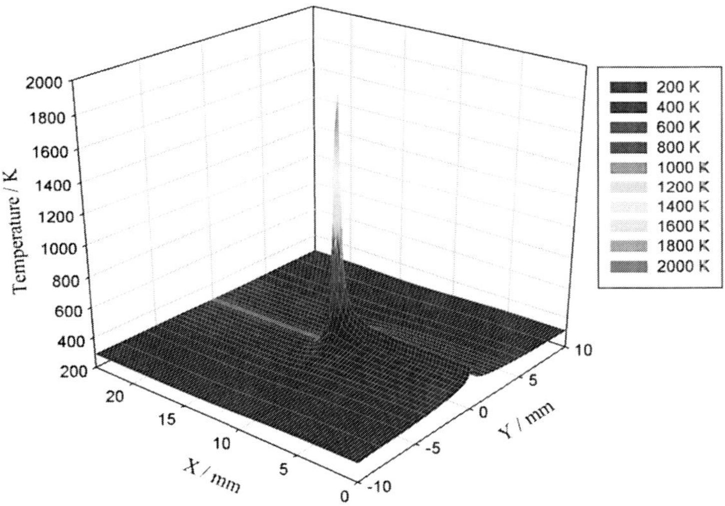

FIGURE 18. Temperature distribution without Ti filler

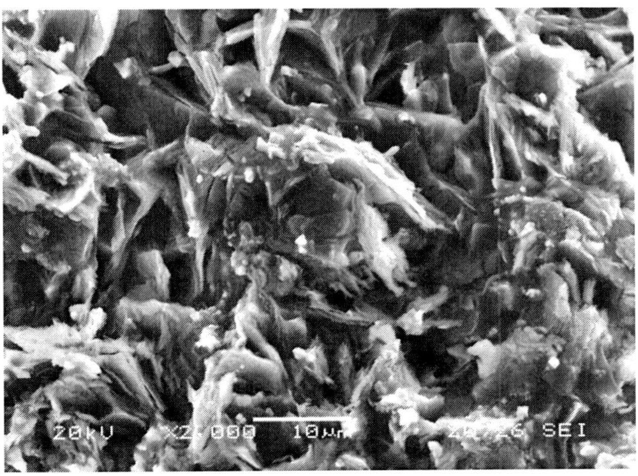

FIGURE 19. Fractograph of the laser welded joint without Ti filler

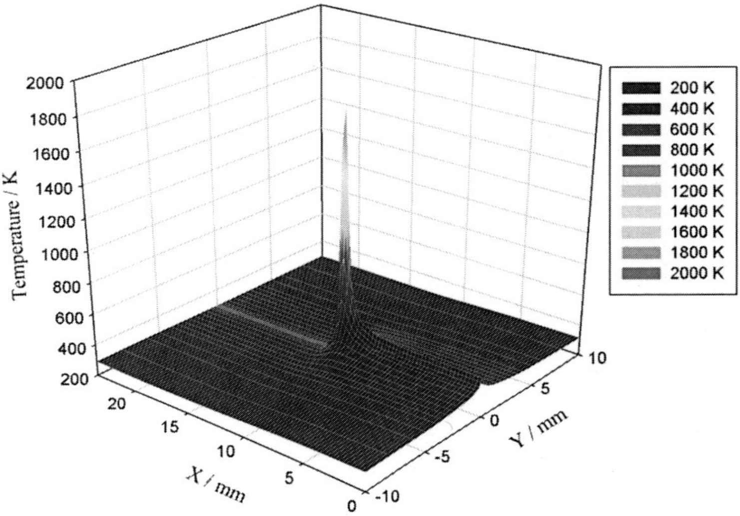

FIGURE 20. Temperature distribution with Ti filler

Figure 20 shows the temperature field of laser welding SiCp/A356 with Ti filler. Considering the Ti melting and in situ reaction in the welding pool as an endothermic reaction, the welding temperature decreases and will be lower than that of laser welding directly (cf. Figs. 18 and 20), and its temperature field is distributed more smoothly with in situ reaction than that of laser welding without Ti filler as shown in Fig. 21. Also, the width of HAZ is decreased to some extent (Fig. 21b). Furthermore, it shows that according to the real effect of laser beam diameter, the thickness of Ti filler is about 0.3 mm will be optimal for in situ welding which conformed to the experimental results as shown in Table 2.

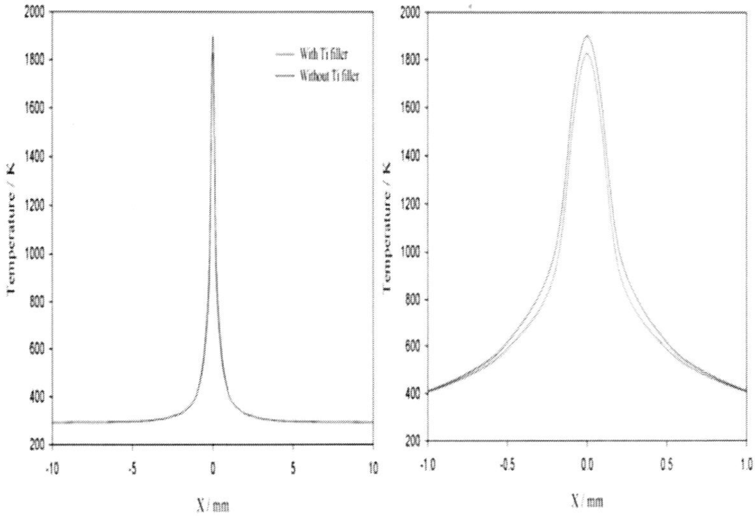

(a) (b)

FIGURE 21. Temperature distribution of central heating on XOZ plane(a) Temperature distribution on XOZ plane (b) Magnification of (a)

In addition, the effect of Ti on the temperature distribution on the central line is shown in Fig. 22. It illustrates that the peak of the temperature is changed distinctly. Because of the sandwiched Ti between the substrates and in situ endothermic reaction, the temperature of substrate ahead of laser resource is lower than that of without Ti filler. Moreover, the temperature at the succedent distance is increased or accumulated a little bit due to the different conductive coefficient between Ti and substrate. On the other side, its corresponding trend of the temperature behind the laser resource (resolidification) is same as that of without Ti filler except for a peak appearance induced by more serious exothermic potential during the crystallization.

Figure 23 shows the temperature distribution when Ti filler is thick. The peak of temperature is decreased obviously and leads to the welding failure.

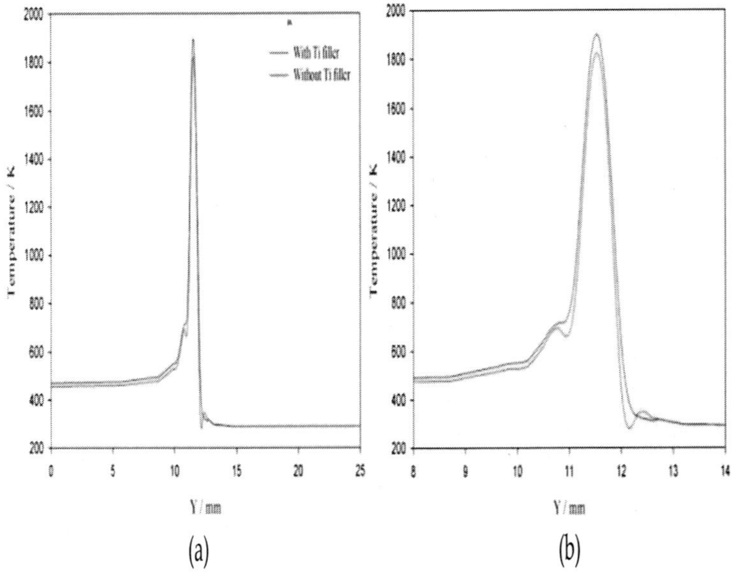

FIGURE 22. Temperature distribution of central line on YOZ plane(a) Temperature distribution on YOZ plane (b) Magnification of (a)

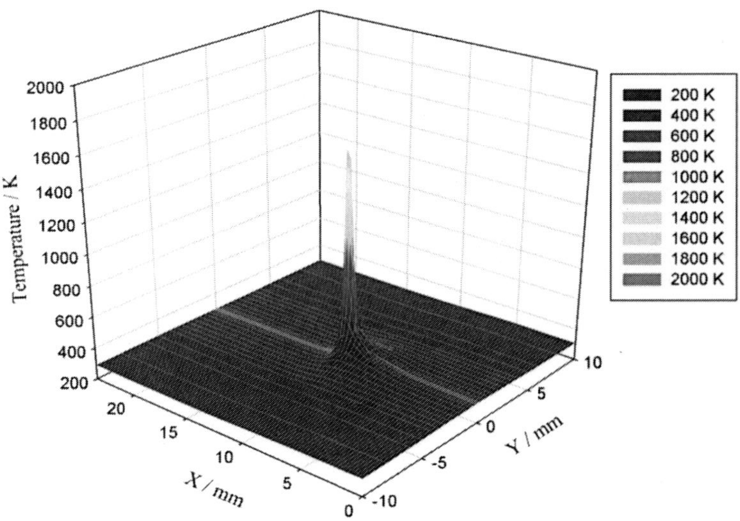

FIGURE 23. Temperature distribution with thick Ti filler

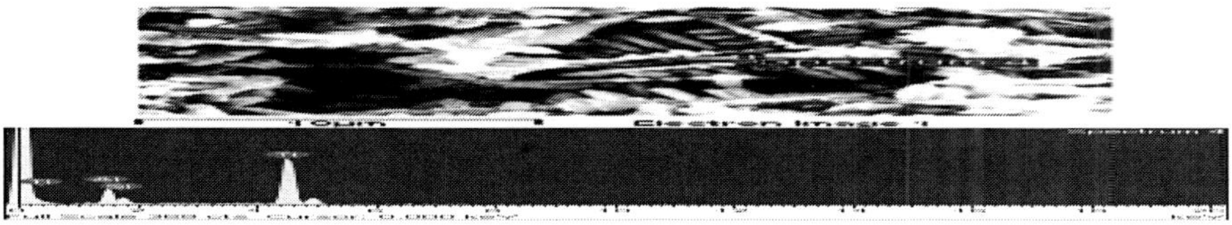

FIGURE 24. Microstructure and EDX of laser weld with thick Ti filler

Figure 24 shows the microstructure of laser welded joint with thick Ti filler and its corresponding energy dispersive X-ray spectroscopy (EDX) results. It can be observed that a large number of columnar Ti crystallization is distributed in the weld. From Figs. 23 and 24, it elucidates that with the increase of Ti thickness, the heat input into the substrate is decreased and most of energy is used for melting Ti led to the insufficient in situ reaction and stirring in the welding pool resulted in lower properties of welded joints.

Furthermore, in order to verify the temperature field, noncontact thermometer (model AZ9881) was used to measure the spot temperature on-line. The measured temperature results are shown in Fig. 25. It shows that the measured results agree well with the simulated results.

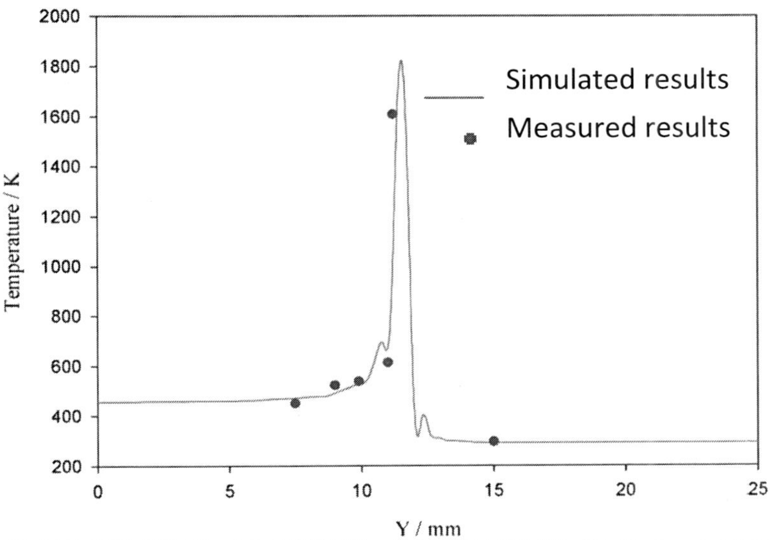

FIGURE 25. Surface temperature distribution in the processing center

4.3. Microstructure Evolution Simulation

According to the temperature calculation, the simulation of the evolution of the microstructure based on thermodynamic equilibria, diffusion [5, 6, 7] was shown in Fig. 26. It showed that during the welding pool solidification, the in situ reinforcement particles TiC would be formed around the initial reinforcement SiC particles. With the increase of cooling time, the initial reinforcement SiC particles would be replaced by the newly-formed in situ reinforcement particles TiC. It was well matched with the results shown in Figs. 7 and 9.

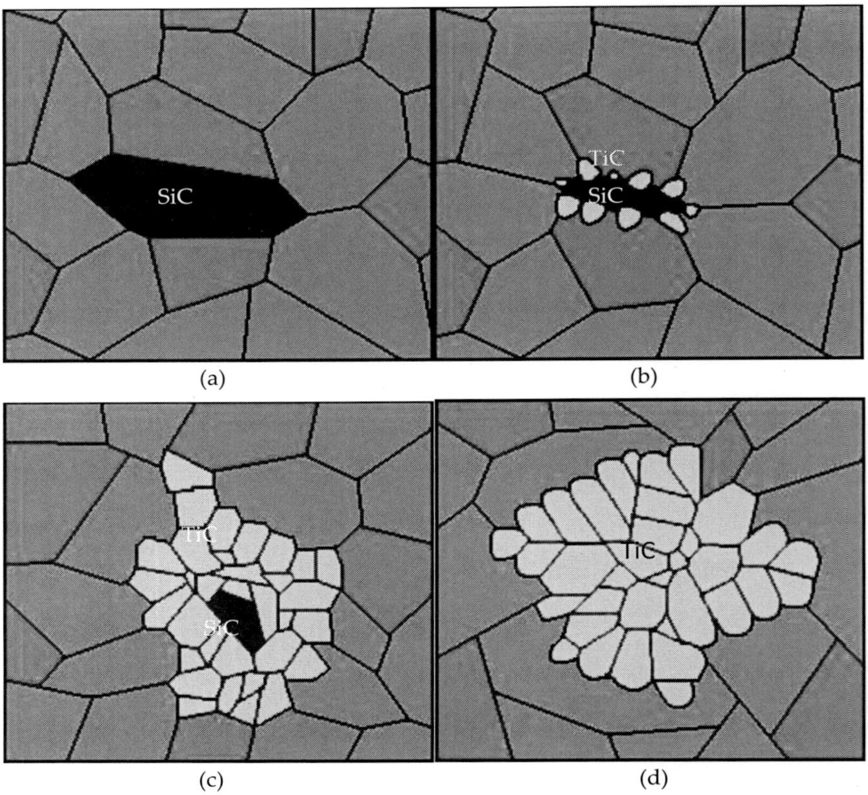

FIGURE 26. Simulation on microstructure evolution(a) Initial (b) Start of solidification (c) Middle of solidification (d) End of solidification

5. CONCLUSIONS

The use of titanium as a filler metal in Nd:YAG laser welding of SiC_p/A356 provided beneficial in situ reinforcement effect. The effect of in situ reinforcement of the Ti filler allowed the newly-formed reinforcement TiC particles to distribute uniformly in the weld that subsequently resulted in successfully welding the SiC_p/A356 composite. Moreover, the typical pernicious interfacial reaction microstructure such as Al_4C_3 was effectively restrained from the interface between aluminum matrix and reinforcement particles in the Nd:YAG laser welding of SiC_p/A356 with Ti filler. Furthermore, according to the temperature calculation, the evolution of the microstructure was simulated based on thermodynamic equilibria, diffusion. Results were well matched with the corresponding experiments.

ACKNOWLEDGEMENT

The work is supported by a RGC general research fund (GRF) (Grant No.:9041503.) and a Strategic Research Grant (SRG) from City University of Hong Kong (Grant No.: 7002287.)

REFERENCES

1 J. M. Go´mez, M. I. de Salazar, Dissimilar. Barrena, Welding. Fusion, A. of, A702, M. M. C. Reinforced, Al. with, 23 Particles. Microstructure and mechanical properties. Materials Science and Engineering A 2003; 352(1-2): 162-168.

2 W. Guo, M. Hua, H. W. Law, J. K. L. Ho, Impact. Liquid-Phase, Welding. Diffusion, Si. of, Cp/60, Al, Mechanism. Its, Materials Science and Engineering: 2008 2008, 490, (1-2), 427 EOF437 EOF .

3 W. Guo, M. Hua, J. K. L. Ho, on. Study, Diffusion. Liquid-Phase-Impact, Si. Welding, Z. L10, Compos. Sci. Technol., 2007 67, (6), 1041 EOF1046 EOF .

4 L. M. Marzoli, A. V. Strombeck, J. F. Dos, C. Santos, L. M. Gambaro, Friction. Volpone, Welding. Stir, an. A. of, A606, 03/20p. Al, Alloy. Reinforced, Composites Science and Technology 2006 66 2 363371 .

5 D. A. Porter, K. E. Easterling, Transformations. Phase, Metals. in, Alloys, 2 Cheltenham: Nelson Thornes, 2001.

6 R. Riedel, of. Handbook, Hard. Ceramic, Materials, New York: Wiley-VCH, Weinheim, 2000

7 R. Boyer, G. Welsch, E. W. Collings, Properties. Materials, Titanium. Handbook, Materials. Alloys, Ohio. A. S. M. Park, International, 1994

8 J. R. Davis, A. S. M. Specialty, Handbook-Aluminum, Alloys. Aluminum, Materials Park, Ohio: ASM International, 1993 557

9 S. Mall, T. Nicholas, Matrix. Titanium, Behavior. Composites-Mechanical, Pa. Lancaster, Pub. Technomic, Co, Inc., 1998

10 W. M. Steen, Material. Laser, Processing, 3 ed. London: Springer-Verlag; 2003.

11 K. W. Guo, of. Influence, Situ. In, on. Reaction, Microstructure. the, Si. of, Si. Al, Welded. Mg, Nd. Y. A. G. by, with. Laser, Filler. Ti, J. Materials Engineering and Performance, 2010 2010 19 5258 .

12 A.F. Mills: Heat and Mass Transfer. P. R. Donnelly & Sons Company; 1995

9

CHAPTER

Development of a Comprehensive Process Model for Hybrid Laser-Arc Welding

Fanrong Kong and Radovan Kovacevic

[1] *Research Center for Advanced Manufacturing, Lyle School of Engineering, Southern Methodist University, Dallas, TX,, USA*

1. INTRODUCTION

Recently, a hybrid welding technique combining laser welding and arc welding has been finding broader applications in industry due to its unique advantages, such as higher welding efficiency and lower costs [1, 2] compared to the traditional arc welding or autogenous laser welding. Because of rapid melting and solidification occurring in the weld zone, a locally high thermal gradient inevitably exists and accompanies with the whole welding process, which really decides the final residual stress and distortion distributions of weld and affects the remained grain size in the fusion zone (FZ) and heat affected zone (HAZ). Also, the levels of residual stresses and distortions directly influence the weld quality [3]. In comparison with traditional arc welding and autogenous laser welding, the temperature field and residual stress distribution in hybrid laser-arc welding involve more variables because of the additional interaction between the laser and arc plasma [4] thus becoming much more complex and difficult to theoretically and experimentally analyze. Trial-and-error experiments are not able to fully describe those physical mechanisms involved in the hybrid laser-arc welding process. Therefore, the numerical tools have been widely used to help explain the complex welding mechanisms present in the hybrid laser-arc welding process [5, 6].

Up to now, numerical work on the welding process mostly concentrates on traditional electric arc welding, including gas tungsten arc welding (GTAW)

[7 - 9], submerged arc welding (SAW) [10,11], gas metal arc welding (GMAW) [12-16], and partly on laser beam welding (LBW) [17-21]. These studies focused on the heat and mass transfer phenomena in the weld pool [8, 11-15], thermal-induced distortion and residual stresses [16, 17, 19, 20], solidification-induced dendrite growth in the FZ [21], and recrystallization in the heat affected zone (HAZ) [9]. Due to the locally rapid melting and solidification occurring in the welding process, a high temperature gradient—which inevitably exists in the weld zone—causes a high-stress concentration in the weld zone and nearby HAZ [22], which usually exceeds the yield strength of the material. Large residual stresses presented in the welded structure can obviously reduce the fatigue strength of metal components, causing crack generation and shorten the lifetime of metal component [23], which could possibly have disastrous results. A number of mitigation procedures to reduce and/or eliminate the level of residual stress have been presented by researchers, mainly including enhancement of the material ductility of solidification zone (SZ) and HAZ, and improving the thermal and mechanical conditions in the welding processes [24].

Because of the complex physical mechanisms in the welding process—which are related to the heat source properties, material performance, and welding parameters, etc.—trial-and-error methods to optimize welding parameters takes a long time and is usually more costly. Also, understanding of the physics of the welding process is limited by only using an experimental approach. Numerical simulation as accompanied by theoretical analysis has been widely applied as a cost-efficient way to help explore the welding phenomena in different welding techniques. Eagar et al. [25, 26] spent a lot of time in developing theoretical models for GTAW processes. Dong et al. [27-29] developed numerical models to predict the residual stresses as well as fatigue life of weld obtained by the multi-pass welding process. Deng et al. [30-32] developed a series of numerical models to study the residual stress distribution in variable welding joints.

Compared to the traditional electric arc welding, laser welding has unique advantages such as high energy density, narrow HAZ, low heat input, and high energy efficiency. However, laser welding is also limited by its disadvantages like poor gap bridgeability and high equipment cost. In order to fully use the advantages of both laser and arc welding techniques, Steen et al. [33] introduced for the first time a hybrid technique by combining the laser beam and arc for welding and cutting in the late 1970s. Subsequently, researcher and engineers have presented a number of works on combining the laser and electric arc in the past decades. Considering that interaction between laser beam and arc plasma is complex, the hybrid laser-arc welding and cladding processes have not been understood fully. Most available literature on these approaches is limited at the level of the experimental study including hybrid laser-GTAW, hybrid laser-GMAW, and hybrid laser-plasma arc welding for steels, magnesium alloy, aluminum alloy, titanium alloy and dissimilar materials. In order to further study the welding mechanism of hybrid laser and arc, it is necessary to develop a comprehensive model to understand the heat and mass transfer, residual stress evolution, as well as microstructure formation in the hybrid laser-arc welding process. Zhou and Tsai [34, 35] presented heat transfer and fluid flow models to study the metal inert gas (MIG) welding and laser-MIG hybrid welding processes. Rao et al. [36] reviewed the modeling of hybrid laser-gas metal arc (GMA) welding and presented further studies on synergistic interaction between the laser beam and arc, the metal transfer features, and behavior of shielding gas. Ribic et al. [37] developed a three-dimensional (3-D) finite volume model to study heat transfer and fluid flow in the hybrid laser-GTA welding process. Considering that the microstructure formation of weld has a close relationship with the macro-scale heat transfer and fluid flow, and residual stress fields, it will be very necessary to integrate the thermal, fluid flow and mechanical modeling with the microstructure evolution like grain growth in the fusion zone and HAZ. Multi-scale and multi-physics modeling is one of most interesting simulation trends in the laser-based heat processes, especially in the hybrid laser-arc welding process.

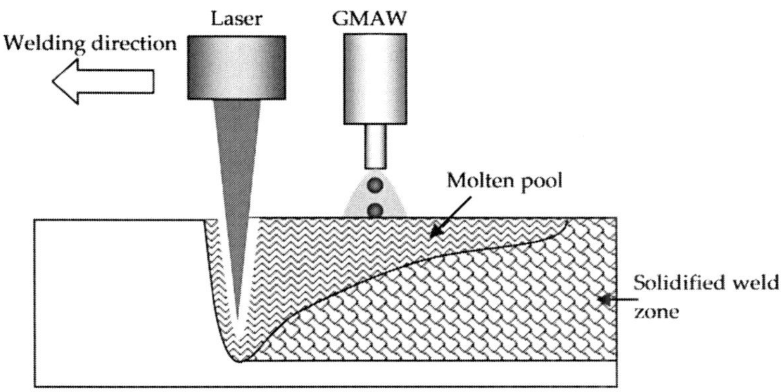

FIGURE 1. Schematic view of hybrid laser-GMA welding system

In this chapter, a 3-D mathematical model will be developed to numerically predict the transient temperature distributions and residual stresses in the hybrid laser-GMA welding of a thick plate of A514 steel in butt joint configuration, as shown in Figure 1. The numerical solution is achieved based on a finite element method by using a commercial numerical package, ANSYS. A Monte Carlo model is introduced to consider the grain growth and phase transformations in the HAZ. The laser and arc heat inputs and heat losses at the surface of coupons are considered by using ANSYS Parametric

Designed Language (APDL). The influences of the processing parameters (including welding speed, laser power, wire feed rate, arc power, and stand-off distance from laser to arc) on the profile and geometrical size of the molten pool, residual stress distribution of the weld, and grain size in the HAZ are numerically studied. The numerically obtained results are experimentally verified.

2. FINITE ELEMENT MODELING

2.1. Thermal analysis of hybrid laser-gma welding

In the hybrid laser-GMA welding process, laser and GMA simultaneously heat the coupon surface in local area, which makes the thermal distribution of weld much more complex. In this study, a cylindrical volume heat-source model with a Gaussian distribution is assumed to simulate the heat input by laser, and a double-ellipsoidal volume heat source is selected to consider the heat input by GMA welding. The general thermal governing equation is shown below, in which thermal conduction-induced heat transfer is considered and temperature-dependent material properties are used [38].

$$\rho c_p \frac{\partial T}{\partial t} = \{L\}^T ([D]\{L\}T) + \dddot{q}_l + \dddot{q}_a \tag{1}$$

$$\text{where}[D] = \begin{bmatrix} k_{xx} & 0 & 0 \\ 0 & k_{yy} & 0 \\ 0 & 0 & k_{zz} \end{bmatrix}, \tag{2}$$

$$\text{and}\{L\} = \begin{Bmatrix} \frac{\partial}{\partial x} \\ \frac{\partial}{\partial y} \\ \frac{\partial}{\partial z} \end{Bmatrix}, \tag{3}$$

ρ is the density, cp is the specific heat, T is the temperature, t is time, kxx, kyy, and kzz are the thermal conductivity components along the x, y, and z axis, respectively; $\dddot{q}_l$ and $\dddot{q}_a$ are the volume heat generation rates due to the laser and GMA heat input, respectively.

So far, a number of heat source models have been developed to simulate the arc welding and laser welding processes. Laser welding usually consists of laser conduction welding or laser keyhole welding. The former one has lower energy density as compared with the latter one by which a keyhole is formed in the weld pool. A surface heat flux model is usually applied in the thermal analysis of laser conduction welding. However, a volume-distributed heat source model, like rotary Gaussian heat density distribution [39], is usually used for simulating a laser keyhole welding. Compared to the laser beam welding, electric arc welding has much lower energy density, and surface heat flux models with Gaussian distributions used to be applied to simulate the arc heat input in the arc welding process.

Considering that the enthalpy brought into the weld pool by melted wire in GMAW, volume-distributed heat source models are preferred, such as hemi-spherical power density distribution [40], ellipsoidal power density distribution [41], and double ellipsoidal power density distribution [42]. However, all of these heat source models are empirically derived based on the experimentally fitting data. Therefore, each heat source model mentioned above has a certain applicable range in the real production case. It is suggested that engineers in the welding process design should reasonably select a heat source model which matches well with the specific welding process. There are a limited number of publications available to numerically describe the hybrid laser and arc welding process because of lack of knowledge on complex interaction between the material, arc plasma, and laser beam [43]. Current heat source models of hybrid laser and arc including GTAW and GMAW were mostly developed with the help of experimental support [34-37].

In this study, a double-ellipsoidal heat source model is introduced to simulate the GMAW heat input, and a cylindrical heat source model with a sectional Gaussian distribution is used to consider the laser heat input. q^f_{arc} (x, y, z, t) and q^r_{arc} (x, y, z, t), depict heat input distributions inside the front and rear quadrants of the GMAW heat source, respectively, which can be expressed as follows [42, 44]:

$$\dot{q}^f_{arc}(x,y,z,t) = \frac{6\sqrt{3} f_f P_{arc}}{abc_f \pi\sqrt{\pi}} \exp\left(-\frac{3(x-x_0)^2}{a^2}\right)\cdot\exp\left(-\frac{3(y-L_w)^2}{b^2}\right)\cdot\exp\left(-\frac{3(z-vt)^2}{c_f^2}\right) \quad (4)$$

$$\dot{q}^{r}_{arc}(x,y,z,t)=\frac{6\sqrt{3}f_r P_{arc}}{abc_r\pi\sqrt{\pi}}\exp\left(-\frac{3(x-x_0)^2}{a^2}\right)\cdot\exp\left(-\frac{3(y-L_w)^2}{b^2}\right)\cdot\exp\left(-\frac{3(z-vt)^2}{c_r^2}\right) \tag{5}$$

where a, b, c_f, c_r are the characteristic parameters of heat sources, and a, b, c_f, and c_r are set at 4 mm, 3 mm, 3 mm and 7 mm respectively [44]. P_{arc} denotes the nominal power of the GMAW, and P_{arc}=μUI. Where μ is the energy efficiency of GMAW based on the welded metal, U denotes the arc voltage of GMAW, and I stands for arc current of GMAW. $\dot{q}_{laser}(x,y,z,t)$ denotes the laser radiation-induced volume heat input given by [45, 46]:

$$\dot{q}_{laser}(x,y,z,t)=\eta_l\frac{P_{laser}\cos\Phi}{2\pi R_l^2}\exp\left(-\frac{(x-x_0)^2+[z+(L_w-y)\tan\Phi-vt-D_{la}]^2\cos^2\Phi}{2R_l^2}\right)\cdot(y/L_w^2) \tag{6}$$

where η_l is laser absorption efficiency based on the welded material, P_{laser} stands for the nominal power of the laser beam, x_0 is the x-coordinate of the center point of laser spot at the coupon surface, L_w is the thickness of the butt joint, and R_l is the effective radius of the laser beam, Φ is the inclination angle of laser head, D_{la} is the laser-to-arc stand-off distance. Φ is set at 0° and D_{la} is set at 8 mm in this study, and *v* denotes the welding speed.

The boundary conditions at the sample surfaces are given by:

$$-k\frac{\partial T}{\partial n}=h_c(T-T_\infty) \tag{7}$$

where *n* is the normal outward vector to the surface of specimen, T_∞ is the room temperature, and *hc*is the heat transfer coefficient of sample surface.

2.2. Mechanical analysis of hybrid laser-gma welding

The mechanical analyses of hybrid laser and arc welding are similar to the previous studies on the electric arc welding and laser welding. The stress and distortion of weld are mainly caused by the thermally-induced expansion and shrinkage and the accompanying phase transformation-induced volume change. By considering the elastic-plastic material properties, stress and strain relationships in the hybrid laser-GMA weld are given by [38]:

$$\{\sigma\} = [D_e]\{\varepsilon^{el}\} \tag{8}$$

where $\{\sigma\}$ denotes the stress vector, $[D_e]$ denotes elastic stiffness matrix, and $\{\varepsilon^{el}\}$denotes elastic strain vector expressed by [30]:

$$\{\varepsilon^{el}\} = \{\varepsilon\} - \{\varepsilon^{th}\} - \{\varepsilon^{pl}\} - \{\varepsilon^{\Delta V}\} - \{\varepsilon^{Trp}\} \tag{9}$$

where $\{\varepsilon\}$ is the total strain vector, $\{\varepsilon^{th}\}$ is thermal strain vector, $\{\varepsilon^{pl}\}$ is the plastic strain vector, $\{\varepsilon^{\Delta V}\}$ is strain vector due to phase transformed induced volume change, and $\{\varepsilon^{Trp}\}$ is strain vector due to phase transformation plasticity which is ignored in this study. The boundary conditions taken into consideration in the mechanical analysis assume that one edge of the butt joint is fixed, and the other one is only transversely shrinkage free.

2.3. Grain size prediction model by monte carlo method

Grain size evolution and phase transformation play a critical role in deciding the final mechanical properties of weld, and it is necessary to involve those factors in the thermo-mechanical modeling of different welding processes. Many good trials have been performed to numerically predict the grain growth in the fusion zone and heat affected zone for solidification and re-crystallization, respectively, which includes Monte Carlo (MC) model [47-50], phase field (PF) method [51], and cellular automaton (CA) model [52] combined with finite element and finite difference analyses. Here a brief introduction of MC model to predict the grain growth in HAZ will be performed. The detailed description of phase field method and cellular automaton model-based numerical prediction of grain growth in welds can be found in literature [53].

Monte Carlo model-based grain growth prediction generally includes the following several steps: (1) The representation of the considered material in a two-dimensional (2-D) or 3-D of cells, as shown inFigure 2a. The content of each cell stands for its crystallographic orientation. A region consisting of a set of consistently distributed cells with the same orientation value denotes a grain. The grain boundaries are identified by a curve in 2-D matrix or a surface in 3-D matrix between the separate planes or volumes with different orientations. (2) After selecting the matrix type and defining it by an initially random number, the free energy of a cell in the matrix with its specific crystallographic orientation based on its surroundings will be identified. (3) Randomly selecting a new crystallographic orientation for each cell. (4) Calculating the free energy of the new coming element with the new crystallographic orientation, the two energy values and their difference are then calculated. A new grain orientation that will minimize the free energy is generated with the selected transition probability [54]. These four steps will be reiterated many times at random positions in the matrix. The ultimate product is a microscopic simulation of the free energy decaying in the system, which is in fact the main driving force for grain growth. The Hamiltonian demonstrates the interaction among the closest neighbors in a particular cell, which stands for the grain boundary energy and can be calculated as follows [54]:

$$G = -J \sum_{nn} (\delta s_i s_j - 1) \tag{10}$$

where, J is a positive constant that characterizes the scale of the grain boundary energy; si is one of the possible orientations, which is set between 1 and q, in the ith cell of the matrix; sj is the crystallographic orientation of one neighboring cell; *nn* is the amount of neighboring cells for each element. In the Monte Carlo model, a Moore neighborhood model is selected (see Figure 2b); therefore, nn=8. Δab is the Kronecker-delta, which equals 1 when two elements in the matrix are equal, i.e., a=b, and 0 for others. As a consequence, neighboring cells with different orientations contribute a free energy *J* to the system and 0 with the same orientation. The total number of grain orientations, *q* is set at 40 in this model since it is known that the grain-growth exponent becomes almost independent of q when its value is larger than 30 [54].

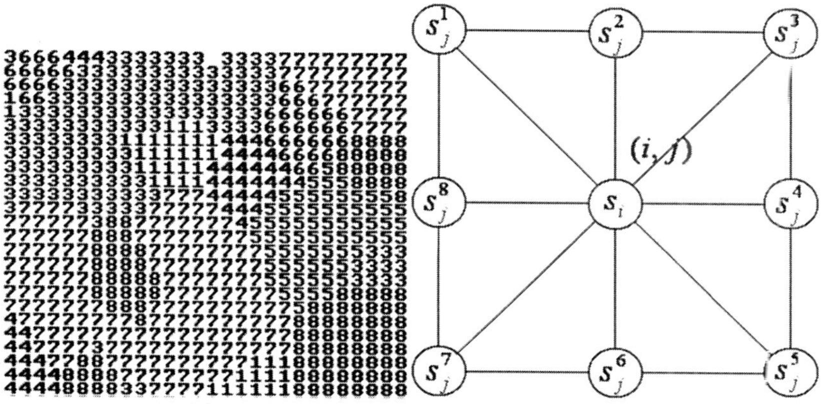

(a) (b)

FIGURE 2. The grain structure in MC model with Moore neighborhood [51](a) The grain structure represented by a 2D square (b) Moore neighborhood

The transition probability W is given by [55]:

$$W = \begin{cases} \exp(-\frac{\Delta G}{k_b T}), & \Delta G > 0 \\ 1, & \Delta G \le 0 \end{cases} \quad (11)$$

where ΔG is the change of the free energy because of the orientation alteration, kb is the Boltzman constant (k_b=1.38×10^{-23} $m^2kg/s^2/K$), and T is the temperature. Therefore, the speed of the moving segment can be calculated by [55]:

$$v_i = C_1[1 - \exp(-\frac{\Delta G}{k_b T})] \quad (12)$$

where C_1 is the boundary mobility. For a continuous grain growth, the final grain size can be calculated by using the following equation [55]:

$$L^n - L_0^n = f(T)t \tag{13}$$

where L and L_0 are the final and initial mean grain sizes respectively calculated by the linear-intercept method, *n* is the grain growth exponent and set at 1.84 in this study [56]. *f*(T) is usually computed as an Arrhenius-type equation [55], and its expression is shown as follows:

$$f(T) = K\exp(-\frac{Q}{R_g T}) \tag{14}$$

where *K* is the pre-exponential coefficient, *Q* is the activation energy for grain growth, and *Rg* is the universal gas constant. In this study, K is set at 3.01×10^{-2}, and Q is set at 1.7×10^{5} J/mol [56].

The Monte Carlo method has been proven to be an effective way to simulate grain growth with slow and uniform temperature evolution such as metal casting [53]. In the hybrid welding by laser and arc, there exists a dynamic thermal process with rapid heating and cooling resulting in an abrupt temperature gradient in the HAZ and fusion zone. In the simulation of microstructure evolution, three techniques—such as the atomistic models, a grain boundary migration (GBM), and experimentally data-based (EDB) models—have been presented [53, 58-60]. The atomistic model used to be only applied to small numbers of atoms like nanocrystals [60], and it is not suitable for a large-scale FZ or HAZ simulation. The GBM model can be a good alternative for grain-growth simulation when the isothermal grain-growth kinetics is not accessible. However, the physical properties of the material in this model have to be known, and the grain size is assumed to be proportional to the square root of time. The EDB model can avoid these shortages and be applied to simulate the grain growth in HAZ when the isothermal grain-growth kinetics of metal are available. Therefore, it can be used to relate time and temperature to the Monte Carlo simulation-time step t_{MCS} [55]:

$$L = K_1\lambda(t_{MCS})^{n_1} \tag{15}$$

where λ is the discrete grid-point spacing in the Monte Carlo model, and K_1 and n_1 are constants. Through the regression computation of t_{MCS} and the Monte Carlo model predicting the grain size, the values of K_1 and n_1 are obtained as 0.715 and 0.477, respectively [56]. In the EDB model, the relationship between the t_{MCS} and the real time-temperature T(t) is further given by [60]:

$$(t_{MCS})^{nn_1} = \left(\frac{L_0}{K_1\lambda}\right)^n + \frac{K}{(K_1\lambda)^n}\sum\left(\exp\left(-\frac{Q}{R_g T(t)}\right)\Delta t_i\right) \tag{16}$$

where *n* is the grain growth exponent, T(t) is the mean temperature in a time interval Δt_i. Therefore, at any given monitoring location where the temperature is known as a function of time, t_{MCS} can be related to the real time t, which is $\sum \Delta t_i$. The t_{MCS} values at different locations calculated through Eq. (16) cannot be straightly applied to the Monte Carlo model since the selection of a grid point for updating the orientation number is stochastic in the Monte Carlo approach. Consequently, the probability of choosing each grid point is the same as in the traditional MC calculations. However, grains usually grow at higher rates in the HAZ region of higher temperature, where a sharp temperature gradient is present. This fact has to be included in any practical grain-growth calculation scheme. One solution is to develop a scheme in which grain orientations at higher-temperature locations (higher t_{MCS} locations) are updated with a higher frequency by considering a probability gradient. In other words, the site-selection probability changes with location. The larger the t_{MCS} at a location, the higher the corresponding site-selection probability [57, 60]:

$$P = t_{MCS} / t_{MCSMAX} \tag{17}$$

where *tMCSMAX* is the maximum of *tMCS* in the simulation domain.

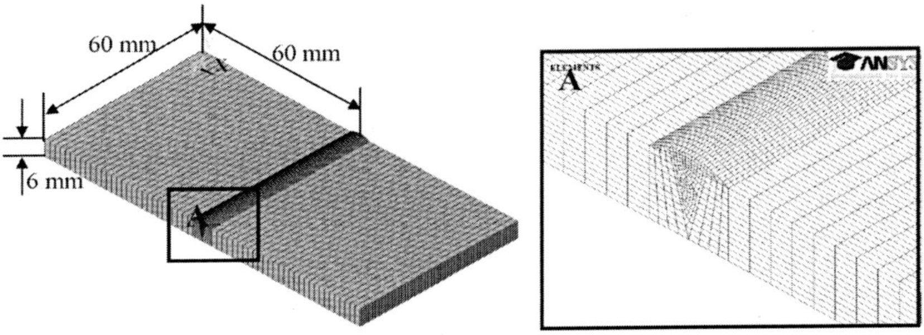

FIGURE 3. Finite element meshes for hybrid laser-GMA weld

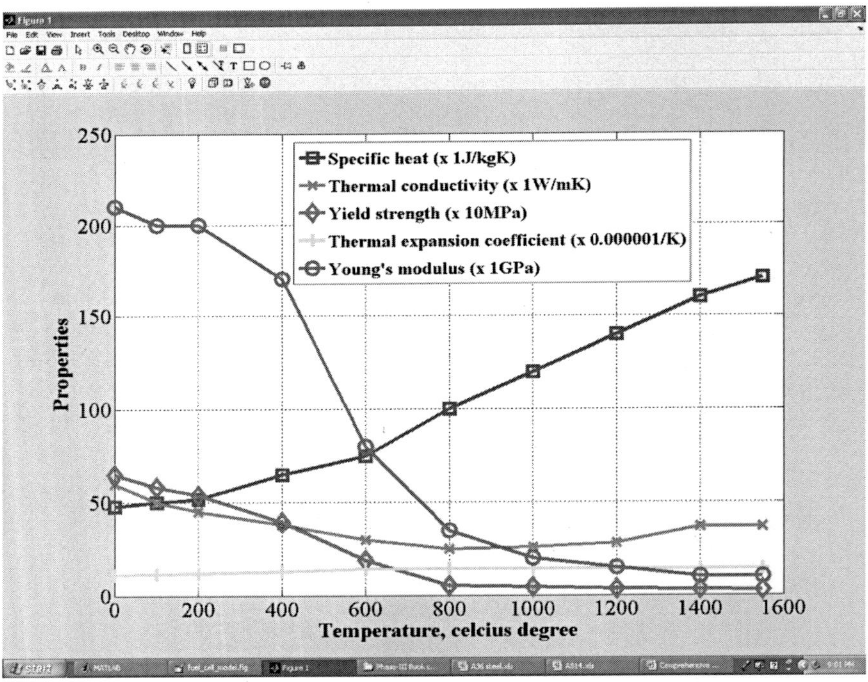

FIGURE 4. Temperature-dependent thermal and mechanical properties of A514 steel

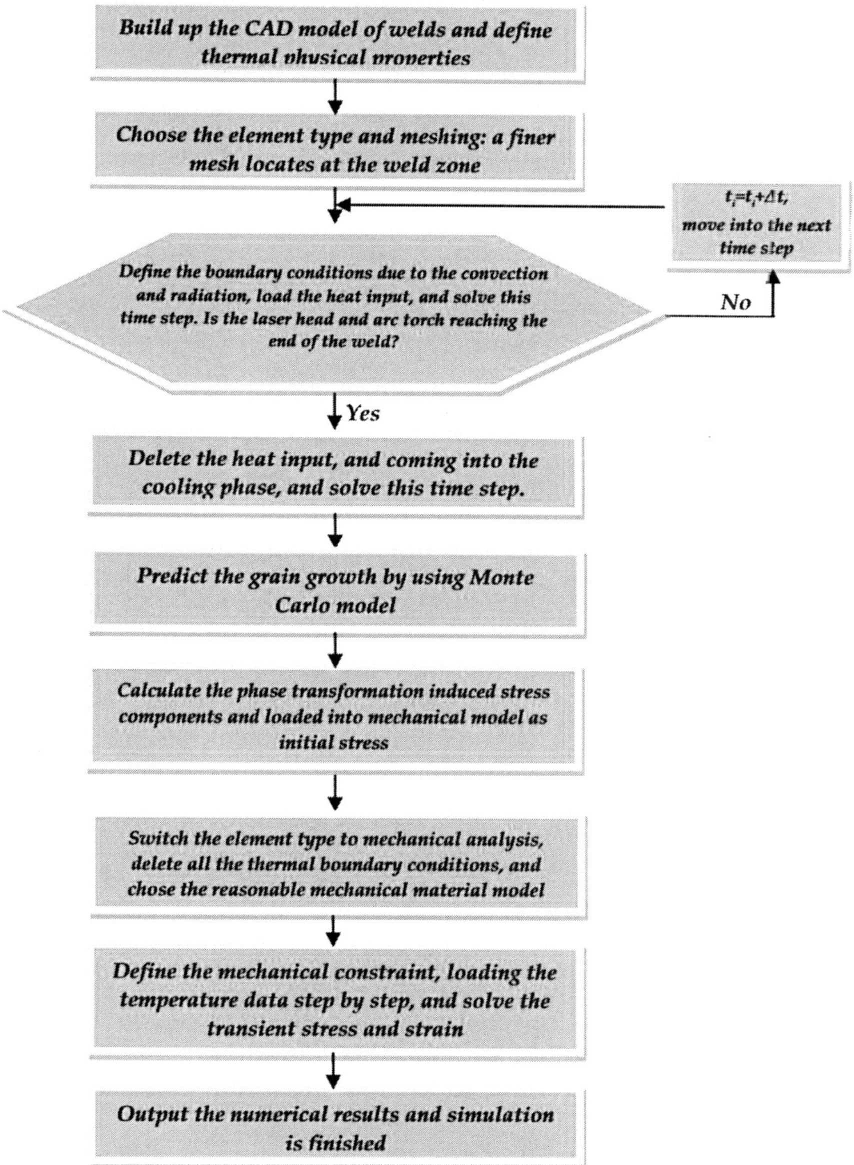

FIGURE 5. Numerical procedure performed in the thermo-mechanical FE analysis

2.4. Implementation of numerical procedure by using apdl

An uncoupled finite element thermo-mechanical model with considering the grain growth in the HAZ by Monte Carlo model is developed to study the temperature distribution and residual stress field in the hybrid laser-GMA welding process. A non-uniform mesh is selected in which a finer mesh is used in the weld bead and a course mesh is defined in the other region of the welded coupons (see Figure 3). A temperature-dependent material property is used in the numerical modeling, as listed in Figure 4. A thermal FE analysis is performed to achieve the temperature field of hybrid laser-GMA welding process. The wire feeding into the groove to form the weld bead has been simulated by using element kill-and-birth approach which is available in ANSYS software. The achieved geometrical size of the weld zone could be compared to the micrographs of the weld cross-section obtained by an optical microscope, by which the accuracy of thermal analysis can be verified. The numerical model is then transferred to mechanical analysis module in ANSYS by switching the element type from thermal to structural. The corresponding constraints are exerted into sample boundaries. The achieved temperature histories are subsequently loaded into the mechanical model step by step to calculate the displacement, stress/strain of the sample due to the thermal expansion or shrinkage during the welding process. A bilinear hardening principle is introduced in this study to simulate the material plastic behavior. Von Mises criterion is used for considering the yield behavior of the sample material. Figure 5 shows the numerical procedure used in this study.

3. EXPERIMENTAL SET-UP

A 4 kW fiber laser and a GMAW torch are mounted on a robotic arm to perform the hybrid welding of a thick plate for a butt joint configuration. The photo of an experimental set-up for welding is shown inFigure 6. In order to control the gap thickness, spot welding is performed at both ends of the joint before the formal welding starts. After the welding process is completed, the achieved sample will be cut into standard tensile coupon for tensile test, the left parts will be mounted for polishing, and etching to test the micro-hardness and microstructure. The base metal is high strength steel A514; its chemical composition is listed in Table 1. The wire material is ER100S-G. Its diameter is 0.9 mm, and its chemical composition is listed in Table 2. Residual stresses were measured by using the X-ray diffraction technique. Before performing the residual stress measurement, the measurement areas were cleaned by using polishing paper.

TABLE 1. Chemical composition of A514 [61]

	C	Mn	P	S	Si	Cr	Mo	V	Ti	B
Min.	0.12	0.70			0.20	0.40	0.15	0.03	0.01	0.0005
Max.	0.21	1.00	0.035	0.008*	0.35	0.65	0.25	0.08	0.04	0.005

TABLE 2. Chemical composition of ER100S-G [62]

Cu % max	Ni % max	Fe % max	Mn % max	Mo % max
<0.5	<5.0	Balance	<5.0	0.50

FIGURE 6. Photo of hybrid laser-GMA welding system used in this study

4. RESULTS AND DISCUSSION

4.1. Experimental results

Figure 7 shows the surface morphology of A514 weld obtained by hybrid laser-GMA welding, andFigure 8 presents the corresponding cross-sectional view of weld. It can be seen that a sound weld quality is achieved by using hybrid laser-GMA welding and the welding-induced cracks can be effectively mitigated by reasonably selecting filler wire matched with base metal. Also, a sound mechanical property can be obtained. Figures 9a and b show the hardness distribution in the weld obtained by hybrid laser-GMA welding.

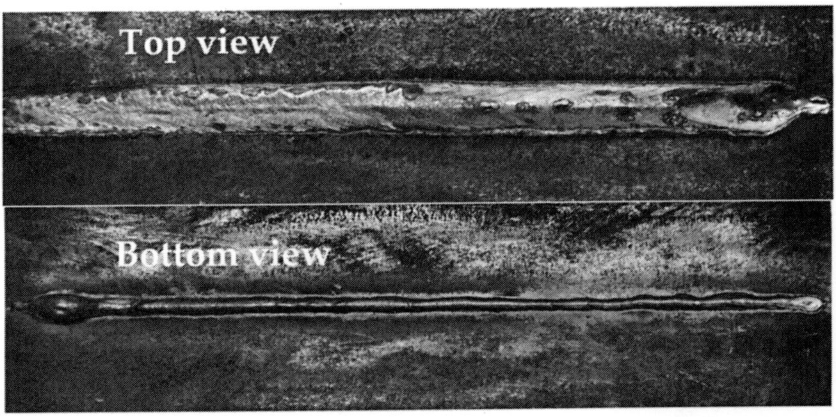

FIGURE 7. A514 sample achieved by 3.8 kW laser and 159A×30.5V GMAW with a welding speed of 12 mm/s and 8-mm stand-off distance between the laser and arc

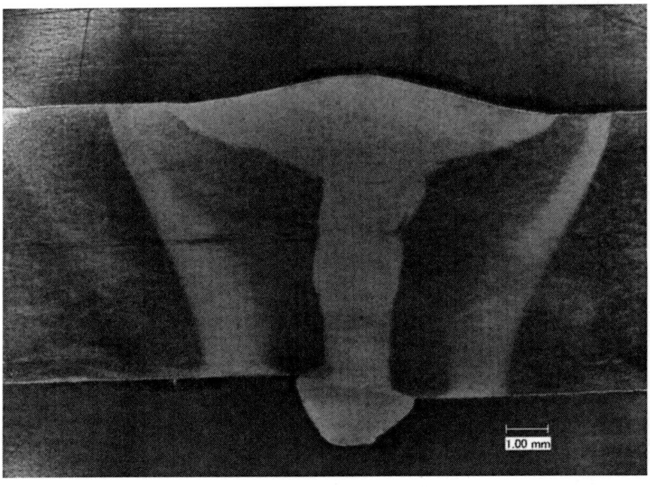

FIGURE 8. Cross-sectional view of A514 weld sample achieved by 3.8 kW laser and 159A×30.5V GMAW with a welding speed of 12 mm/s and 8-mm stand-off distance between the laser and arc

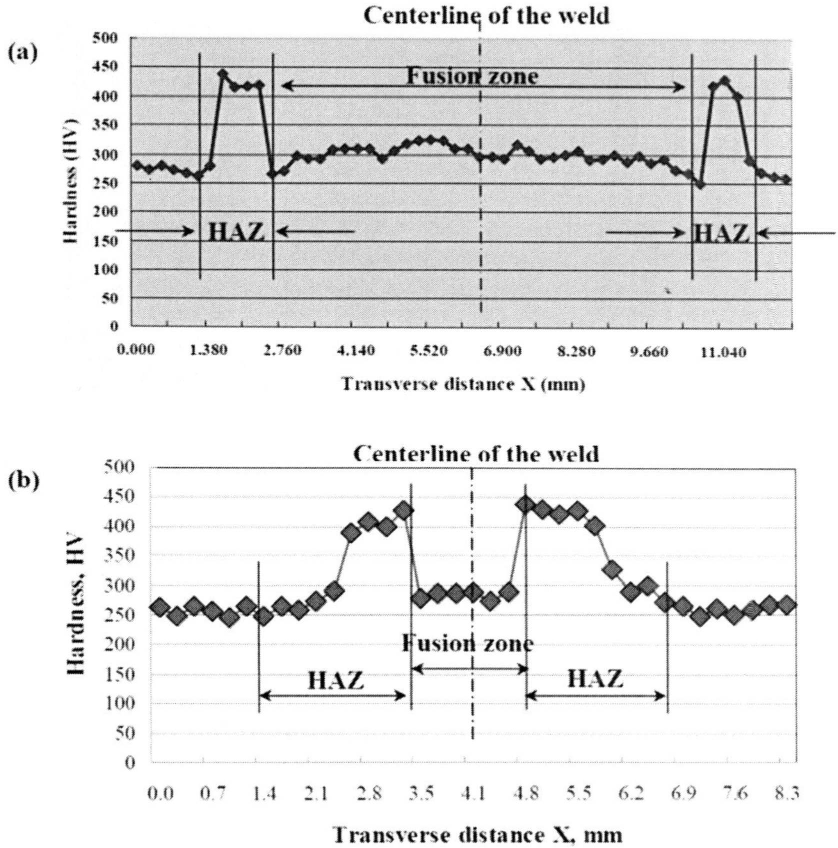

FIGURE 9. Hardness distribution transverse to the weld in A514 sample, (a) at the top surface and (b) at the bottom surface

4.2. Thermal results and grain size prediction in the haz of hybrid laser-gma welding

Finite element analyses results show the temperature at the each location of weld with respect to the welding time. The temperature evolution curves at position A, B, and C are shown in Figure 10, where position A is located at the center of weld, position B is at the heat-affected zone, and position C is in the base metal, as shown in Figure 11b. Figures 11a and b show the top and cross-sectional views of weld obtained by hybrid laser-GMA welding, respectively. It is inevitable that material heating and cooling is accompanied by phase transformation and grain size change, especially in the HAZ of weld, which is the weakest zone of the weld. In this study, a Monte Carlo-based sub-model is introduced to numerically predict the grain growth in the HAZ combined with the finite element thermal analysis.Figure 12a shows the relationship of temperature versus Monte Carlo step at Position B, and Figure 12b presents the curve of Monte Carlo step versus real time at Position B. The corresponding predicted grain size distribution in the Position B is shown in Figure 12c. The numerically predicted grain size is compared to grain size shown in the micrograph of the cross-section of weld (see Figure 12d), and a qualitative agreement is achieved.

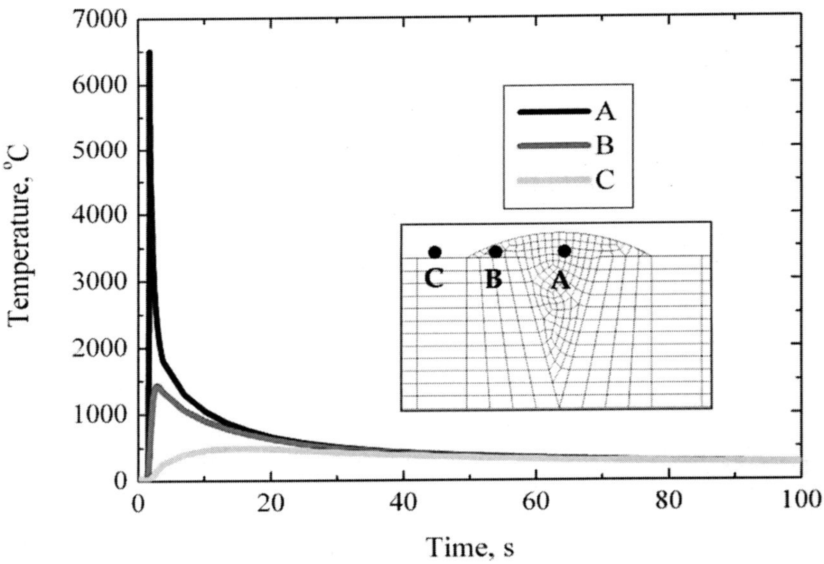

FIGURE 10. Temperature evolution curve at the FZ, HAZ and BM during the hybrid laser-GMA welding of A514 steel

4.3. Thermally-induced residual stress in the hybrid laser-gma welding

A finite element analysis is further performed to predict the thermally-induced residual stress distribution based on the previous thermal analysis results. The contours of transient stress, along thickness normal stress, longitudinal stress and equivalent residual stress of hybrid weld are shown

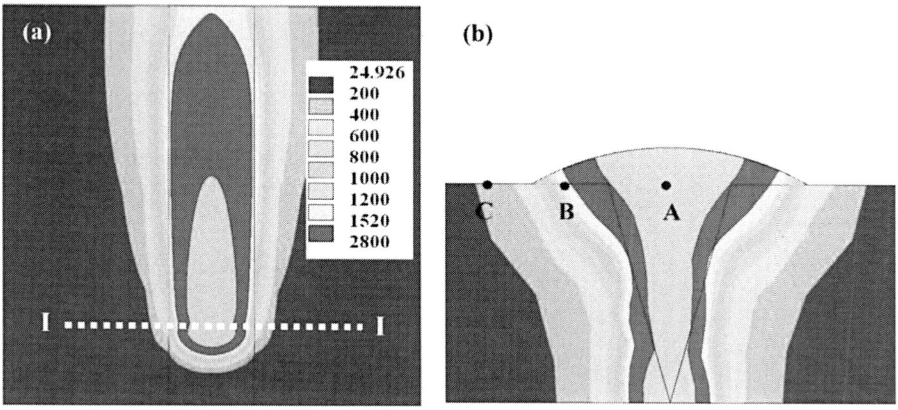

FIGURE 11. Numerically predicted isotherms at the top (a) and at the cross-section I-I of the weld (b) obtained by hybrid laser-GMA welding

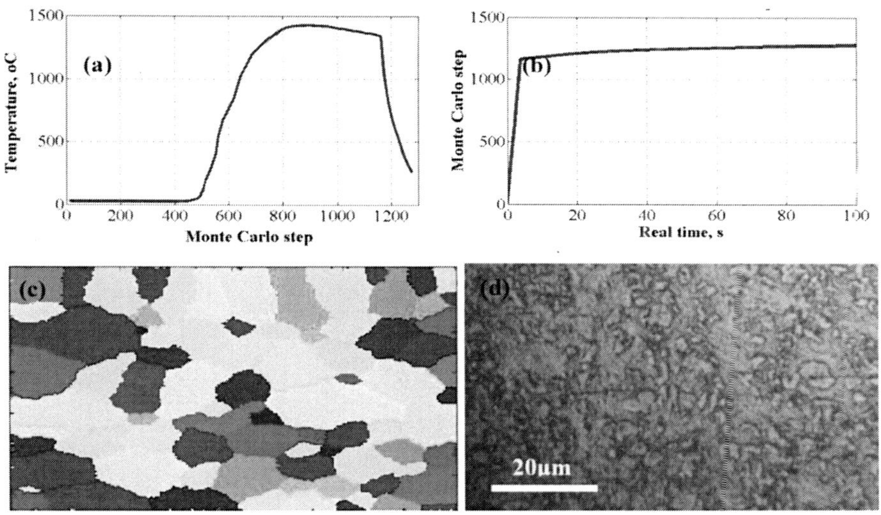

FIGURE 12. a) Temperature versus Monte Carlo step, (b) Monte Carlo step versus real time, (c) numerical predicted and (d) experimentally measured grain size distributions at Position B in the HAZ of hybrid A514 weld by Monte Carlo sub-model

In Figures 13a through d, respectively. It can be seen that the higher stress concentrations are located at the weld zone, which also indirectly verifiesthe previously experimentally obtained conclusions that the thermally induced cracks are usually generated at the weld zone, not in the base metal. The corresponding contours of stress distribution of the cross-section in the middle of weld length are shown in Figures 14a through d. A higher stress concentration is found to be located at the top region of cross-section

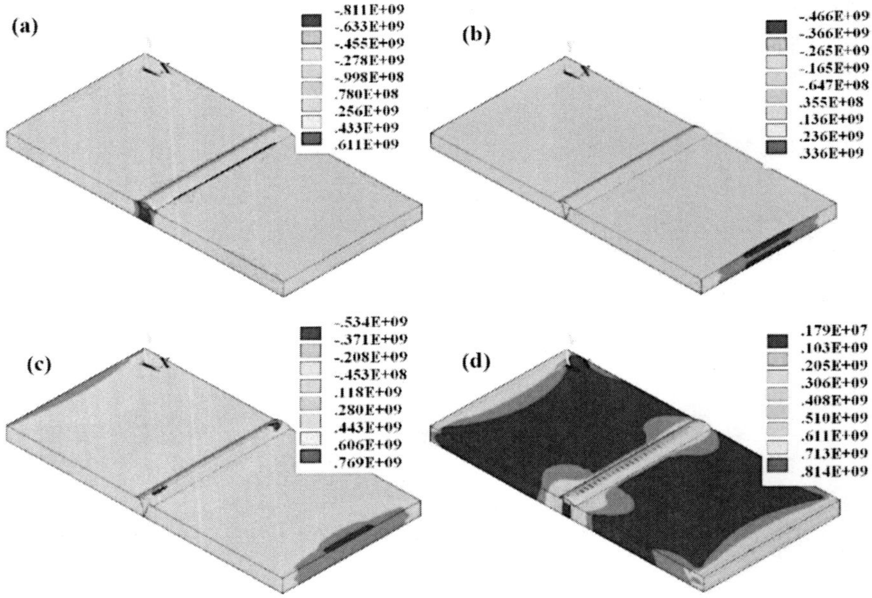

FIGURE 13. a) Transverse stress SX, (b) along-thickness normal stress SY, (c) longitudinal stress SZ, and (d) von Mises equivalent residual stress SEQV mapping of weld by hybrid laser-GMA welding (unit of stress in the contour is Pa)

Figures 15a through c show residual stress distribution transverse to the weld bead at the different thicknesses in the middle of weld obtained by hybrid laser-GMA welding. It also validates the conclusion driven from Figure 14 that high tensile transverse and longitudinal stresses are located at the top and bottom regions of the weld center, high compressive transverse stresses are located at a half of the weld thickness. From the equivalent stress distribution point of view, the peak value of stress concentration is a little lower than that at the top and bottom of the weld. Figure 16 shows residual stress distribution transverse to the weld bead at the different locations along the top surface of weld obtained by hybrid laser-GMA welding. Figure 17 also shows residual stress distributions along the central line at the top surface of weld centerline achieved by hybrid laser-GMA welding. It is clear that stress distribution across the weld bead is uniform along the weld; only a little drop in stress magnitude exists at the both ends of weld.

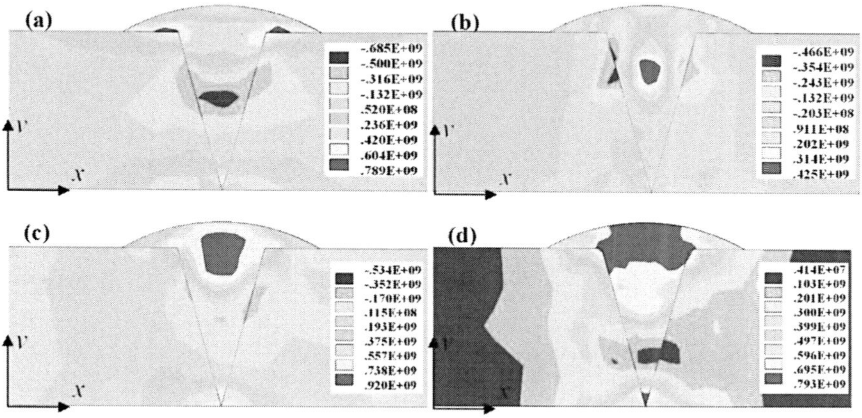

FIGURE 14. a) Transverse stress SX, (b) along-thickness normal stress SY, (c) longitudinal stress SZ, and (d) von Mises equivalent residual stress SEQV mapping of cross-section of weld by hybrid laser-GMA welding (unit of stress in the contour is Pa)

Figure 18 shows a comparison of experimentally-measured and FE numerically-predicted residual stress distributions at the middle of weld length of top surface of weld by hybrid laser-GMA welding. There is a qualitative agreement between the developed numerical model and experimentally measured stress by an X-ray diffraction technique. Figures 19, 20 and 21 show the transverse, longitudinal, and equivalent stresses as well as temperature evolution with time at the positions A, B and C, respectively. It can be seen that the peak values of transient stresses at positions A, B, and C are sensitive to the temperature curve at the position A but not at the positions B and C.

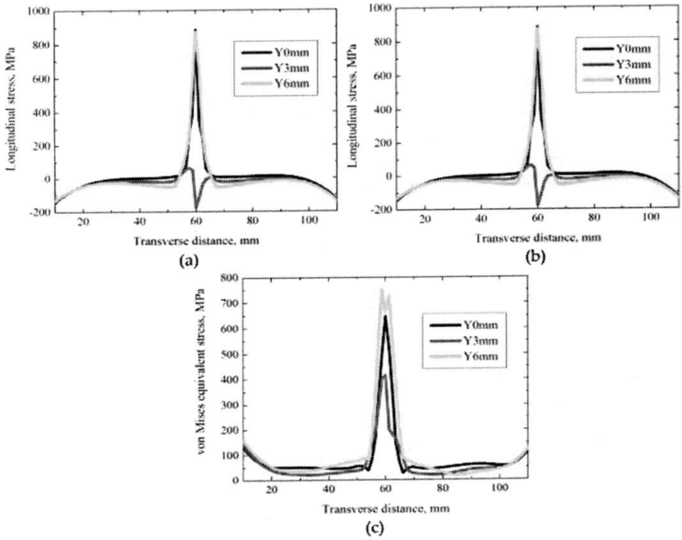

FIGURE 15. Residual stress distribution transverse to the weld bead at the different thicknesses in the middle of weld (z=30 mm) obtained by hybrid laser-GMA welding

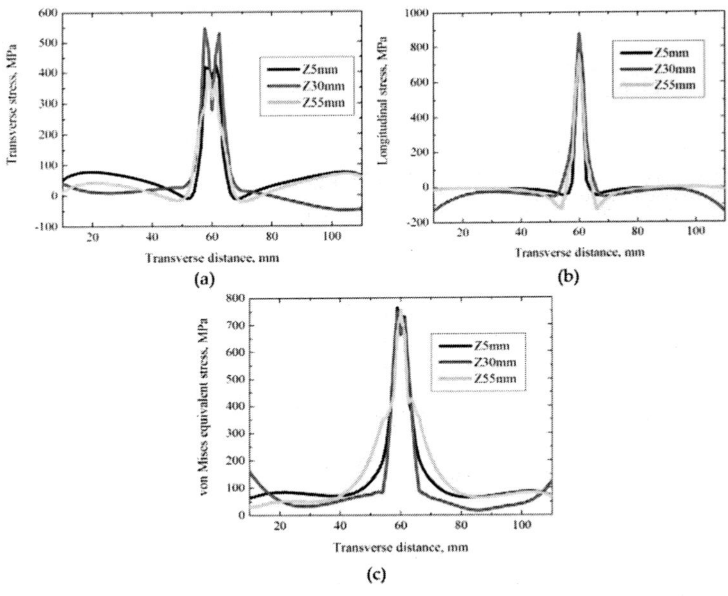

FIGURE 16. Residual stress distribution transverse to the weld bead at the different locations along the top surface of weld obtained by hybrid laser-GMA welding

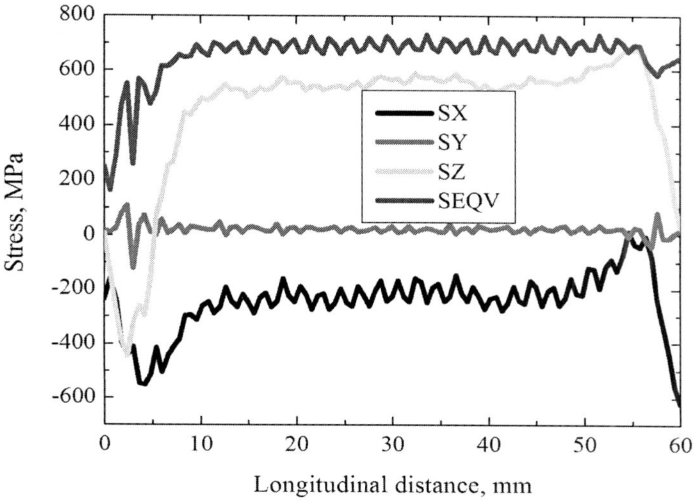

FIGURE 17. Residual stress distributions along the longitudinal direction of weld at the top surface of weld centerline achieved by hybrid laser-GMA welding

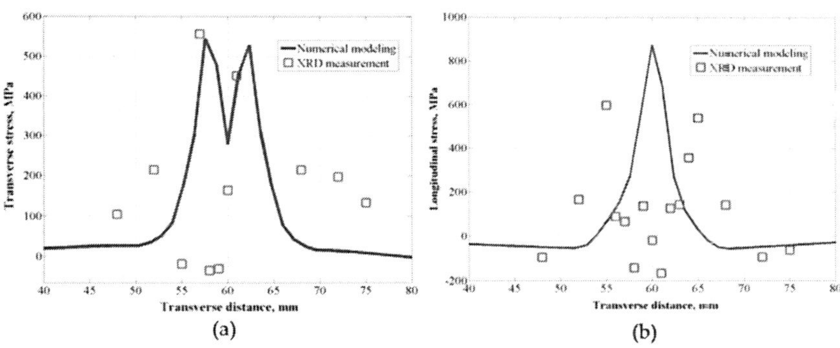

FIGURE 18. Comparison of experimentally-measured and FE numerically-predicted residual stress distribution at the middle of weld length of top surface of weld by hybrid laser-GMA welding

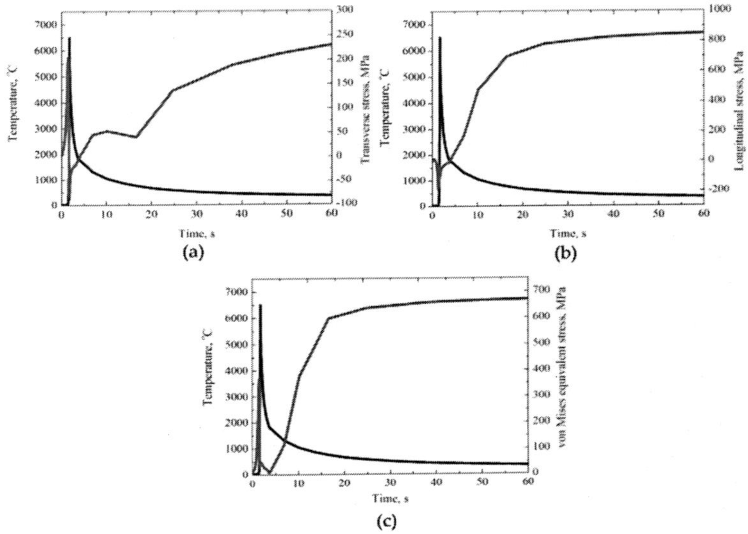

FIGURE 19. Stress and temperature evolution curves at the Position A in FZ of hybrid A514 weld

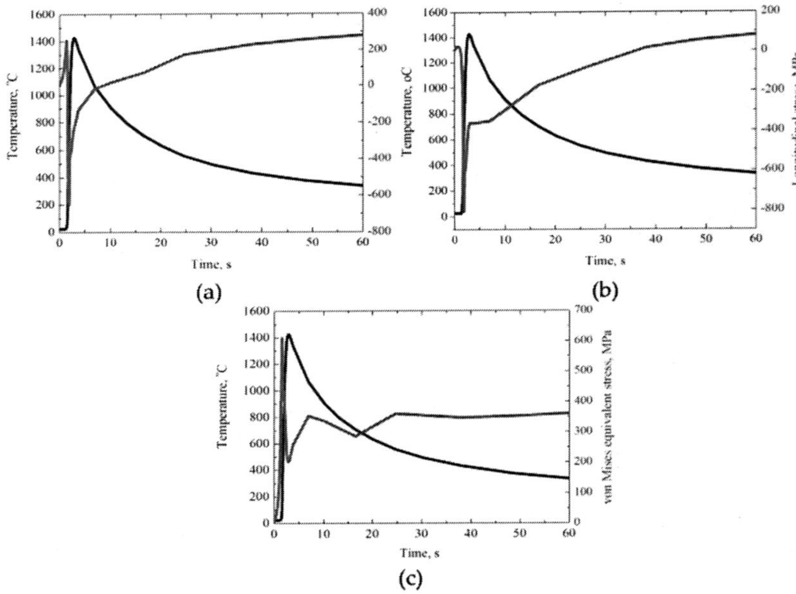

FIGURE 20. Stress and temperature evolution curves at the Position B in HAZ of hybrid A514 weld

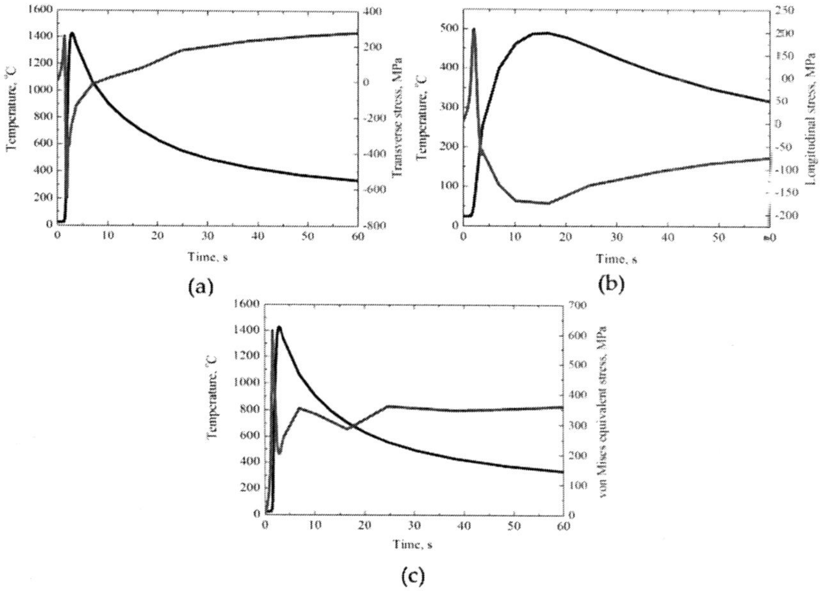

FIGURE 21. Stress and temperature evolution curves at the Position C in BM of hybrid A514 weld

5. CONCLUSIONS

The hybrid approach combining laser and arc has unique features which definitely help to achieve a better weld quality and to improve the production efficiency. A brief overview of modeling of hybrid laser-arc welding process has been presented in which heat transfer, fluid flow, residual stress and distortion, as well as phase transformation in the weld zone and heat affected zone, are involved. As a case study, a 3D thermo-mechanical finite element model is developed to study the thermally-induced residual stress in the hybrid laser-GMA welding process. A Monte Carlo model is introduced to numerically predict the grain growth in the heat affected zone of weld combined with finite element thermal analysis, which can be used to further understand the welding mechanisms of hybrid laser-GMA welding as well as other welding technology.

ACKNOWLEDGEMENT

The authors would like to thank Ph.D. candidate, Mr. Junjie Ma for his valuable help in the experiment and technical discussions and Research Engineer, Mr. Andrew Socha for his support in the experimental set-up. This work was financially support by NSF's Grant No. IIP-1034652

REFERENCES

1 C. Roepke, S. Liu, S. Kelly, R. Martukanitz, Hybrid laser arc welding process evaluati36 on DH36 and EH36 steel. Welding Journal, 2010, 89: 140s-150s.

2 C. Bagger, F. W. Olsen, Review of laser hybrid welding. Journal of Laser Applications. 2 EOF14 EOF .

3 M. N. James, D. J. Hughes, Z. Chen, H. Lombard, D. G. Hattingh, D. Asquith, J. R. Yates, P. J. Webster, stresses. Residual, performance. fatigue, Failure. Engineering, Analysis, 200714 384 EOF395 EOF -385.

4 P. Kah, Usability of laser-arc hybrid welding processes in industrial applications. Ph.D. Dissertation, Lappeenranta University of Technology, 2011

5 G. Turichin, E. Valdaytseva, I. Tzibulsky, A. Lopota, O. Velichko, Simulation and technology of hybrid welding of thick steel parts with high power fiber laser. Physics Procedia, 2011 12: 646 EOF655 EOF .

6 M.H. Cho and D.F. Farson. Simulation study of a hybrid process for the prevention of weld bead hump formation. Welding Journal, 2007 86: 174s-179s.

7 F. G. Lu, S. Yao, S. N. Lou, Y. B. Li, Modeling and finite element analysis on GTAW arc and weld pool. Computational Materials Science, 2004 29:371 EOF -378.

8 H. G. Fan, H. L. Tsai, S. J. Na, Heat transfer and fluid flow in a partially or fully penetrated weld pool in gas tungsten arc welding. International Journal of Heat and Mass Transfer, 2001 44: 417 EOF428 EOF .

9 H. J. Aval, S. Serajzadeh, A. H. Kokabi, Prediction 304 grain growth behavior in HAZ during gas tungsten arc welding of 304 stainless steel. Journal of Materials Engineering and Performance, 2009, 18: 1193 EOF1200 EOF .

10 L. J. Yang, M. J. Bibby, R. S. Chandel, Linear regression equations for modeling the submerged arc welding process. Journal of Materials Processing Technology, 1993 39(1-2): 33 EOF42 EOF .

11 D. V. Kiran, B. Basu, A. K. Shah, S. Mishra, A. De , Three-dimensional heat transfer analysis of two wire tandem submerged arc welding. ISIJ International, 2011 51 (5): 793 EOF798 EOF .

12 J. Haidar, A. theoretical, for. model, metal. gas, welding. arc, tungsten. gas, welding. I. arc, Journal of Applied Physics, 1998 84: 3518-3529.

13 J. Haidar, Prediction of metal droplet formation in gas metal arc welding II. Journal of Applied Physics, 1998 84: 3530 EOF3540 EOF .

14 J. Haidar, An analysis of heat transfer and fume production in gas metal arc welding. III, Journal of Applied Physics, 1999 1999 853448 .

15 Z. H. Rao, J. Hu, S. M. Liao, H. L. Tsai, Modeling of the transport phenomena in GMAW using argon-helium mixtures. Part I- The arc. International Journal of Heat and Mass Transfer, 2010 53: 5707 EOF5721 EOF .

16 C. Heinze, C. Schwenk, M. Rethmeier, Numerical calculation of residual stress development of multi-pass gas metal arc welding. Journal of Constructional Steel Research, 12 EOF19 EOF .

17 A. Kaplan, A. model, deep. of, laser. penetration, based. welding, calculation. on, the. of, profile. keyhole, Journal of Physics D: Applied Physics, 1994 27(9): 1805 EOF -1814.

18 R. Spina, L. Tricarico, G. Basile, T. Sibillano, Thermo-mechanical modeling of laser welding of 5083 sheets. Journal of Materials Processing Technology, 2007, 191: 215 EOF219 EOF .

19 A. P. Mackwood, R. C. Crafer, Thermal modeling of laser welding and related processes: a literature review. Optics & Laser Technology, 2005 37: 99 EOF115 EOF .

20 P. Martinson, S. Daneshpour, M. Koçak, S. Riekehr, P. Staron, Residual stress analysis of laser spot welding of steel sheets. Materials & Design, 2009 30(9): 3351 EOF3359 EOF .

21 W. Tan, N. S. Bailey, Y. C. Shin, A. novel, model. integrated, cellular. combining, automata, field. phase, for. methods, evolution. microstructure, solidification. during, multi-component. of, alloys. multi-phase, Computational Materials Science, 2011 50: 2573 EOF2585 EOF .

22 K. Asim, J. Lee, J. Pan, Failure mode of laser welds in lap-shear specimens of high strength low alloy (HSLA) steel sheets. Fatigue & Fracture of Engineering Materials & Structures. 2011 35: 219 EOF237 EOF -236.

23 Y. Arai, M. Kikuchi, T. Watanabe, M. Nakagaki, Residual stress due to welding and its effect on the assessment of cracks near the weld interface. International Journal of Pressure Vessels and Piping. 1995 63(3): 237 EOF -248.

24 Y. P. Yang, P. Dong, J. Zhang, X. Tian, A. hot-cracking, technique. mitigation, welding. for, aluminum. high-strength, alloy, Welding Journal, 2000 79 EOFs EOF -17s.

25 A. H. Dilawari, T. W. Eagar, J. Szekely, An analysis of heat and fluid flow phenomena in electroslag welding. Welding Journal, 1978 57(1): 24s-30s.

26 L. A. Jones, T. W. Eager, J. H. Lang, Magnetic forces acting on molten drops in gas metal arc welding, Journal of Physics D: Applied Physics, 1998 31: 93 EOF -106.

27 P. Dong, Residual stress analyses of a multi-pass girth weld: 3 -D special shell versus axisymmetric models. ASME Journal of Pressure Vessel Technology, 2001 123: 207-213.

28 P. Dong, J. K. Hong, P. Rogers, Analysis of residual stresses in Al-Li repair welds and mitigation techniques, Welding Journal (London), 1998 77 (11): 439-445.

29 P. Dong, J. K. Hong, J. Zhang, P. Rogers, J. Bynum, S. Shah, of. Effects, weld. repair, stresses. residual, wide-panel. on, loaded. specimens, tension. A. S. M. in, ASME Journal of Pressure Vessel Technology, 1998 120: 122 EOF -128.

30 Dean Deng, Yu Luo, Hisashi Serizawa, Masakazu Shibahara, and Hidekazu Murakawa. Numerical simulation of residual stress and deformation considering phase transformation effect. Transactions of Joining and Welding Research Institute (Osaka University) 2003 32(2): 325-333.

31 D. Deng, H. Murakawa, Numerical simulation of temperature field and residual stress in multi-pass welds in stainless steel pipe and comparison with experimental measurements. Computational Materials Science, 2006 37: 269 EOF277 EOF .

32 D. Deng, F. E. M. prediction, welding. of, stress. residual, in. distortion, steel. carbon, phase. considering, effects. transformation, Materials and Design, 2009 30: 359 EOF366 EOF .

33 W. M. Steen, M. Eboo, Arc augmented laser welding. Metal Construction, 1979 11(7): 332-335.

34 J. Zhou, H. L. Tsai, Modeling of transport phenomena in hybrid laser-MIG keyhole welding. International Journal of Heat and Mass Transfer, 2008 51: 4353 EOF4366 EOF .

35 J. Hu, H. L. Tsai, Heat and mass transfer in gas metal arc welding. Part I: The arc. International Journal of Heat and Mass Transfer, 2007 50 (5-6): 833 EOF846 EOF .

36 Z. H. Rao, S. M. Liao, H. L. Tsai, Modelling of hybrid laser-GMA welding: review and challenges. Science and Technology of Welding and Joining, 2011 16 (4): 300 EOF305 EOF .

37 B. Ribic, R. Rai, T. Deb, Roy, Numerical simulation of heat transfer and fluid flow in GTA/laser hybrid welding. Science and Technology of Welding and Joining. 2008 13 (9): 683 EOF693 EOF .

38 A. N. S. Y. S. Inc, A. N. S. Y. S. , 110 Manual, 2007.

39 S. Wu, H. Zhao, Y. Wang, X. Zhang, A. new, source. heat, in. model, simulation. numerical, high. of, beam. energy, welding, Transaction of China Welding Institute, 2004 25: 91 EOF94 EOF .

40 V. Pavelic, R. Tanbakuchi, O. A. Uyehara, P. S. Myers, Experimental and computed temperature histories in gas tungsten arc welding of thin plate, Welding Journal Research Supplement, 1969 48: 259s-305s.

41 E. A. Bonifaz, Finite element analysis of heat flow in single-pass arc welds. Welding research supplement, 2000 May: 121s-125s.

42 J. Goldak, A. Chakravarti, M. Bibby, A. new, element. finite, for. model, sources. heat, Transaction. Metallurgical, 1984 1984, 15B: 299305 .

43 P. Peyffarth, I. V. Krivt, Welding and Allied Processes Volume I. Laser-Arc Processes and Their Application in Welding and Material Treatment. Tailor&Francis, London, 2002

44 F. Kong, J. R. Kovacevic, Numerical and experimental study of thermally induced residual stress in the hybrid laser-GMA welding process. Journal of Materials Processing Technology, 2011 211(6): 1102 EOF1111 EOF .

45 C. S. Wu, Q. X. Hu, J. Q. Gao, An adaptive heat source model for finite element analysis of keyhole plasma arc welding. Computational Materials Science. 2009 46: 167 EOF172 EOF .

46 F. Kong, R. Kovacevic, . D. finite, modeling. element, the. of, induced. thermally, stress. residual, the. in, laser/arc. hybrid, of. welding, joint. lap, Journal of Materials Processing Technology, 2010 210 (6-7): 941 EOF950 EOF .

47 M. Nosonovsky, X. Zhang, S. K. Esche, Scaling of Monte Carlo simulation of grain growth in metals. Modelling and Simulation in Materials Science and Engineering, 2009 17: 025004 EOF (13pp).

48 S. Mishra, T. Debroy, Measurements and Monte Carlo simulation of grain growth in the heat-affected zone of Ti-6Al-4V welds. Acta Materialia 2004 52: 1183 EOF1192 EOF .

49 S. Sista, Z. Yang, T. Debroy, Three-dimensional Monte Carlo simulation of grain growth in the heat-affected zone of a 2.25Cr-1Mo steel weld. Metallurgical and Materials Transactions 2000 2000, 31B 529 EOF536 EOF .

50 F. Kong, S. Santhanakrishnan, D. Lin, R. Kovacevic, Modeling of temperature field and grain growth of a dual-phase steel 980 in direct diode laser heat treatment. Journal of Materials Processing Technology, 2009, 209 (18-19): 5996 EOF -6003.

51 A. Farzadi, M. -Quang, S. Serajzadeh, A. H. Kokabi, G. Amberg, Phase field simulation of weld solidification microstructure in an Al-Cu alloy. Modelling and Simulation in Materials Science and Engineering, 2008 16: 165005 EOF (18pp).

52 Ch.A. Grandin, J.L. Desbiolles, M. Rappaz, and Ph. Thevoz. A three-dimensional cellular automaton-finite element model for the prediction of solidification grain structures. Metallurgical and Materials Transaction A, 1999, 30A: 3153-3165.

53 Dierk Raabe, Franz Roters, Frédéric Barlat, Long-Qing Chen, Continuum Scale Simulation of Engineering Materials: Fundamentals-Mcrostructures-Process Applications, Wiley-VCH, Weinheim, 2004.

54 B. Radhakrishnan, T. Zacharia, Simulation of curvature-driven grain growth by using a modified monte carlo algorithm. Metallurgical and Materials Transactions 1996 1996, 26(1): 167 EOF -180.

55 S. Sista, Z. Yang, T. Debroy, Monte. Three-dimensional, simulation. Carlo, grain. of, in. growth, heat-affected. the, of. a. zone, 2., Cr, steel. Mo, Metallurgical. weld, Transactions. Materials, 2000 2000, 31: 529 EOF536 EOF .

56 S. Mishra, T. Debroy, Measurements, Carlo. Monte, of. simulation, growth. grain, the. in, zone. heat-affected, Ti. of, Al, V. welds, Materialia. Acta, 2004 52: 1183 EOF1192 EOF .

57 Y.H. Wei, Y.L. Xu, Z.B. Dong, and J.L. Xia, Three dimensional Monte Carlo simulation of grain growth in HAZ of stainless steel 316 Key Engineering Materials, 2007, 353-358: 1923 EOF1926 EOF .

58 M. , Y. Li, E. Kannatey-Asibu, Carlo. Monte, of. simulation, zone. heat-affected, in. microstructure, welded. laser-beam, sheet. nickel, Journal. Welding, , 2002 2002 3744 .

59 Weibin Wang, Yaowu Shi, Yongping Lei, and Zhiling Tian, FEM simulation on microstructure of DC flash butt welding for an ultra-fine grain steel, Journal of Materials Processing Technology, 2005,161: 497-503.

60 J. H. Gao, R. G. Thompson, time-temperature. Real, for. models, Carlo. Monte, of. simulations, grain. normal, Acta. growth, Materialia, 1996 44: 4565 EOF4570 EOF -4575.

61 http://www.ssab.com/Global/SSAB/SSAB_Americas/A514%20Grade%20S%20Brochure%2006-25-08.pdf

62 http://www.midalloy.com/html/pdf/low_alloy/TB%20ER100S-G%20Weld.pdf

11

CHAPTER

Nd:YAG Laser Welding for Photonics Devices Packaging

Ikhwan Naim Md Nawi[1], Jalil Ali[2], Mohamed Fadhali[3] and Preecha P. Yupapin[4]

[1] Centre of Foundation Studies & Faculty of Applied Sciences, Universiti Teknologi MARA, Malaysia

[2] Universiti Teknologi, Malaysia

[3] Physics Department, Faculty of Science, Ibb University, Yemen

[4] King Mongkut's Institute of Technology Ladkrabang, Bangkok, Thailand

1. INTRODUCTION

The state-of-the art of the pulsed Nd:YAG laser spot welding for photonics device packaging has been introduced by Marley (Marley, 2002), which utilizes the laser for high precision joining and alignment. The advantages of laser welding over conventional fusion welding processes include precise welds with a high aspect ratio, narrow heat affected zone (HAZ), very little thermal distortion, ease of automation, high welding speed, enhanced design flexibility, clean, high energy density, low heat input and an efficient process (Zhou & Tsai, 2008; Kazemi & Goldak, 2009). One of the key features of laser welding is the ability to weld without filler materials and it offers distinct advantages (Pang et al., 2008). Laser welding is a liquid-phase fusion process. It joins metals by melting the interfaces and resulting the mixing of liquid molten metal. Then, it solidifies on the removal of the laser beam irradiation (Ready, 1997). Photonics devices used for telecommunications of military applications are usually required to operate for a long life of operation in fields with potentially humid, corrosive and mechanically turbulent environments. Therefore, long term reliability in such hostile operating conditions requires strong fixing of the aligned components and hermetic sealing of the photonics devices inside metal hybrid housings (Fadhali et al., 2007a). It is worth to mention that 60% to 80% of the photonics devices modules cost are due to the coupling and packaging processes. Therefore, understanding the effective packaging technique is very important to produce efficient and reliably packaged photonics devices. Moreover, care must be taken to assemble functional packages because only packages that can be manufactured reliably at competitive costs will survive in the business world. For photonics packaging applications, most of the welds are of butt or lap joints which require the weld penetration depth to be larger than the bead width. Moreover, for miniature packages that contain some sensitive coupling components, the penetration depth should be large enough to achieve a strong attachment (Fadhali et al., 2007a, 2007b, 2007c, 2007d, 2008). At the same time, the bead width should be small to minimize the HAZ and hence prevent the damage to the sensitive optical components. The desired material for this application requires a low thermal conductivity or a higher electrical resistivity (Dawes, 1992). The lower the thermal conductivity of a material the more likely it is to absorb laser energy. For this reason, several weldable grades of steel and stainless steel are ideal for laser welding. The low carbon austenitic stainless steel (300 series steel) which has carbon level less than 0.1% produces good quality welds and reliable weld performance (Fadhali et al., 2007a; Naim et al., 2009). Hence, in this chapter, a stainless steel 304 is utilized as a base material for laser welding. It has been reported that other types of stainless steel were studied by other researchers. For instance, Mousavi and Sufizadeh investigated stainless steel 321 and 630 (Mousavi & Sufizadeh, 2008), while Beretta et al. studied stainless steel 420 (Beretta et al., 2007), for the application of pulsed Nd:YAG laser welding. Whilst a great deal of effort has been focused on developing processing systems there is an urgent need to understand the strength of the weldment. It is the aim of this chapter to examine the strength of a stainless steel 304 welded joint. Despite the pulsed Nd:YAG laser welding has been widely used in microelectronic and photonics packaging industry, a full understanding of various phenomena involved is still a matter of trials and speculations.

2. MATHEMATICAL MODEL OF PENETRATION DEPTH OF A SPOT WELD

Penetration depth of laser spot welding is one of the vital parameters that contributes to a good laser spot micro welding outcome. Practical penetration depth measurements are time consuming and laborious like cutting the samples in order to obtain the weld cross section. Thus, this section shows a theoretical model to estimate penetration depth inside the welded samples by controlling laser parameters or to predict the laser parameters at the required penetration depth. Figure 1 illustrates the penetration depth on a welding specimen. The 1-Dimensional model is developed base on the heat conduction and energy balance equations (Naim et al., 2010).

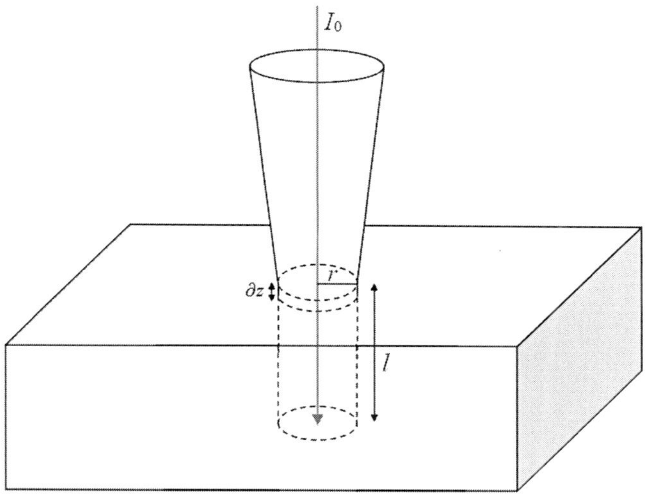

FIGURE 1. Illustration of the penetration depth of the focused laser beam on a welding specimen.

Applying the energy balance equation at the laser spot gives (Naim et al., 2010),

$$(1-R)I_0 = \rho L_m \frac{\partial l}{\partial t} - K\left(\frac{\partial T}{\partial z}\right)_{z=0} \tag{1}$$

Here I_0 is laser power density, L_m is latent heat of fusion, R is material reflectivity, ρ is material density, K is thermal conductivity, $\partial T/\partial z$ is temperature gradient at the welding front, *l* is the penetration depth, *t* is irradiation time and $\partial l/\partial t$

is penetration velocity. To determine the penetration velocity which is the change of penetration depth with time, the temperature gradient at the welding front should first be determined. The temperature distribution inside the target material is governed by the 1-Dimension heat conduction equation given as (Naim et al., 2010),

$$\frac{1}{\alpha}\frac{\partial T}{\partial t}=\frac{\partial^2 T}{\partial z^2} \tag{2}$$

where, α is thermal diffusivity. Equation (2) can be rewritten as

$$-\frac{1}{\alpha}\frac{\partial l_p}{\partial t}\frac{dT}{dz}=\frac{d^2 T}{dz^2} \tag{3}$$

The boundary conditions are taken as,

$$z=0,\ T=T_m \text{ and at } z\rightarrow\infty,\ T=T_0\ . \qquad ()$$

The maximum attainable temperature is the melting temperature of the material, *Tm*. Substituting these boundary conditions into Equation (3) and the solution for temperature distribution inside the material as

$$\frac{T-T_0}{T_m-T_0}=\exp\left[-\frac{1}{\alpha}\left(\frac{dl}{dt}\right)z\right] \tag{4}$$

The temperature gradient at the welding front can be obtained using Equation (4),

$$\left(\frac{dT}{dz}\right)_{z=0} = -\frac{1}{\alpha}\left(\frac{dl}{dt}\right)\left(T_m - T_0\right) \tag{5}$$

Substituting Equation (5) into the energy balance equation,

$$\left(1-R\right)I_0 = \rho L_m\left(\frac{dl}{dt}\right) + \rho c\frac{dl}{dt}\left(T_m - T_0\right) \tag{6}$$

From Equation (6), the welding velocity can be rewritten as,

$$\frac{dl}{dt} = \frac{\left(1-R\right)I_0}{\rho\left[L_m + c\left(T_m - T_0\right)\right]} \tag{7}$$

Integrating Equation (7) gives,

$$l = \frac{\left(1-R\right)Pt}{\pi r_o^2 \rho\left[L_m + c\left(T_m - T_0\right)\right]} \tag{8}$$

Here P is the laser peak power, t is the irradiation time or pulse duration and r is the laser beam radius. From Equation (8), one can observe that penetration depth is proportional to laser peak power and pulse duration as well as the material reflectivity.

Figure 2 illustrates the penetration depth versus peak power and pulse duration. Any changes of either peak power or pulse duration gives the same influence to the penetration depth. This is in agreement with the fact that the penetration depth increases linearly with both the peak powers and pulse durations. The penetration depth increases linearly with an increase of laser peak power. The higher laser peak power means more laser energy is absorbed into the welding material. When a higher laser energy is absorbed, it will produce a deeper melting pool. After solidification, the melting pool produces deeper weld or penetration depth. The trend is similar for variation of pulse duration. Higher pulse duration means longer heating duration. This will provide more time for laser beam to penetrate into the welding material. Thus, a deeper penetration depth will be produced. From Figure 3, it can be summarized that the penetration depth decreases significantly with laser spot radius. This is because the laser power density, I is proportional to $1/r^2$ and the power density will drop significantly as the laser spot radius increases. Laser power density determines the quantity of energy applied on welding material at a specific time. Higher power density will produce a deeper penetration depth.

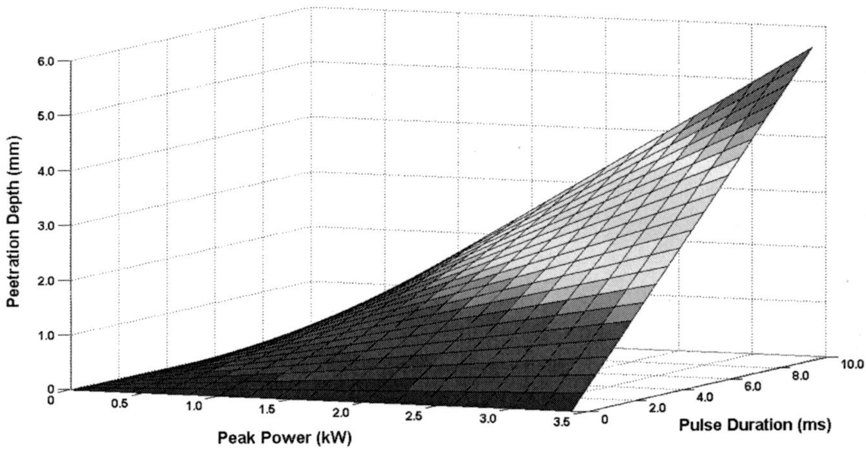

FIGURE 2. Penetration depth versus peak power and pulse duration.

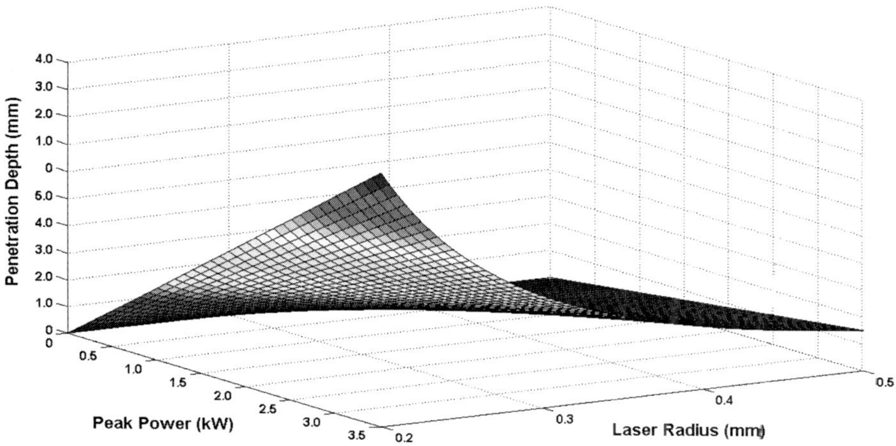

FIGURE 3. Penetration depth versus peak power and laser beam radius.

3. MATHEMATICAL MODEL OF LASER BEAM POWER PENETRATION

In order to produce a spot weld, a laser beam is incident on a specimen surface. The incident laser beam will be absorbed by the specimen depending on the material absorption coefficient. Thus, the absorbed laser beam heats up the specimen surface, raising its temperature. It will melt the specimen and produce a weld when it solidifies. The laser beam penetration in welding material has been derived using the continuity equation (Naim et al., 2010). It is assumed that laser beam penetrates the material based on a cylindrical coordinate. Figure 4 illustrates the laser beam power penetration in stainless steel specimen.

The laser beam power penetration is derived from a continuity equation given as (Naim, 2010),

$$\frac{\partial P}{\partial t} - k\nabla P = 0 \tag{9}$$

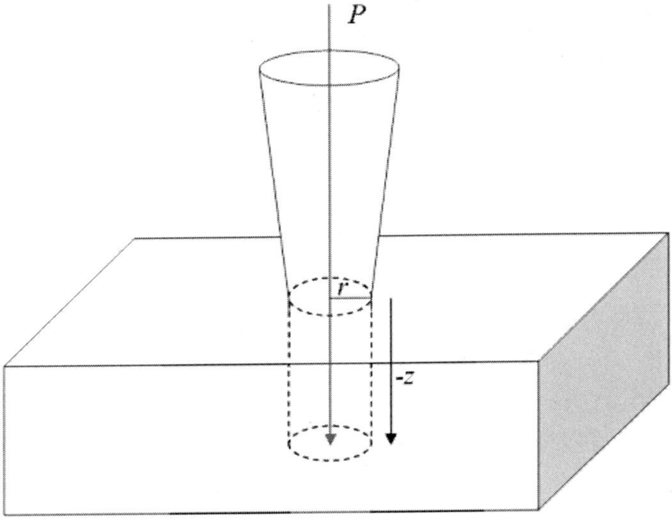

FIGURE 4. Laser beam power penetration in stainless steel specimen.

In cylindrical coordinate, Equation (9) then becomes,

$$\frac{\partial P}{\partial t} = k\left(\frac{\partial^2 P}{\partial r^2} + \frac{1}{r}\frac{\partial P}{\partial r} + \frac{\partial^2 P}{\partial z^2}\right) \tag{10}$$

Here $P=P(r,z,t)$ is the laser beam power penetration, r is the laser spot radius, z is the depth, k is the thermal conductivity and t is time. To solve this equation, let $P = RZT = R(r)Z(z)T(t)$ and the solutions are given by,

$$T = c_1 e^{-\mu^2 kt}, \tag{11}$$

$$R = c_2 J_0(\mu r) + c_3 Y_0(\mu r), \tag{12}$$

and

$$\text{and} \quad Z = c_4 e^{\mu z} + c_5 e^{-\mu z} \tag{13}$$

where *c1, c2, c3, c4, c5* and μ are the unknown constants and determined by the boundary conditions, J_0 is first order of Bessel function and Y_0 is second order of Bessel function. Equation (10) can be solved by using Equations (11), (12) and (13) giving,

$$P(r,z,t) = [c_1 e^{-\mu^2 kt}][c_2 J_0(\mu r) + c_3 Y_0(\mu r)][c_4 e^{\mu z} + c_5 e^{-\mu z}] \tag{14}$$

Let $A = c_1 = c_2 = c_3 = c_4 = c_5$ as the incident laser beam peak power and μ is the material absorption coefficient. In this calculation, only first order of Bessel function is considered. It is also assumed that the laser beam propagates only directing into the material which is –*z* direction. Equation (14) can be written as,

$$P(r,z,t) = Ae^{-k\mu^2 t}[J_0(\mu r)][e^{-\mu z}] \tag{15}$$

For stainless steel material, the absorption coefficient, μ is 0.3 (Kazemi & Goldak, 2009) and the laser beam peak power used in this consideration is 3.5kW. Equation (15) then specified as,

$$P(r,z,t) = 3500e^{-0.3^2 kt}[J_0(0.3r)][e^{-0.3z}] \tag{16}$$

Equation (16) provides the time dependent laser beam penetration. The laser beam penetration is computed for its radius and depth. Figure 5 shows the profile of applied peak power versus the depth of specimen and laser spot radius. The peak power 3.500kW decreases exponentially with depth of stainless steel material and decreases according to the first order of Bessel function in terms of the spot radius. The peak power incident on the surface is 3.500kW at the centre of laser spot target, but decreases to 3.0248 kW at the spot radius of 0.25mm. Peak power reduces to 0.1743kW at the centre and 0.1506kW at spot radius of 0.25mm for a depth of 1.00mm. At the laser spot radius of 2.00mm, the laser beam is almost fully penetrated with only peak power of 0.0087kW at the centre of the penetration.

Figure 6 illustrates the time dependent laser beam penetration for stainless steel material. According to the illustration, the peak power reduces exponentially for both depth and time. From observation, laser beam peak power decreases faster through the depth rather than through the time, relatively. This is explained by Equation (16) which states penetrated laser beam is based on absorption coefficient through the depth is -0.03 m but only -0.0009 m through time. At a depth of 2.00 mm and time of 4.0 ms, the peak power is 0.0002 kW. At this point, laser beam is almost fully penetrated. On the interface between stainless steel material and laser beam spot, laser beam penetration can be illustrated byFigure 7 at certain time. As discussed before, laser beam penetration decreases according to the first order of Bessel function with the spot radius and exponentially with time. At the central of the interface, the peak power is 3.5 kW and it reduces to 0.0026 kW after 8.0 ms. At laser spot radius of 0.25 mm, the laser beam penetration is 3.0248 kW and it reduces to 0.0023 kW after 8.0 ms where the laser beam is about to fully absorbed by the stainless steel material.

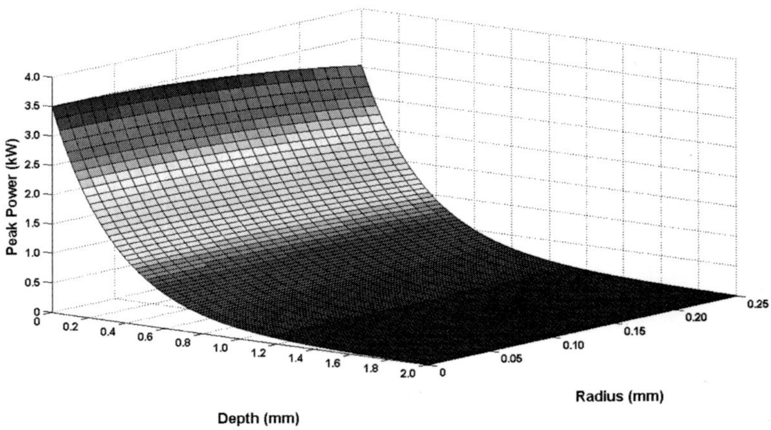

FIGURE 5. Peak power versus depth and radius for stainless steel with absorption coefficient, =0.3 and t=1 ms.

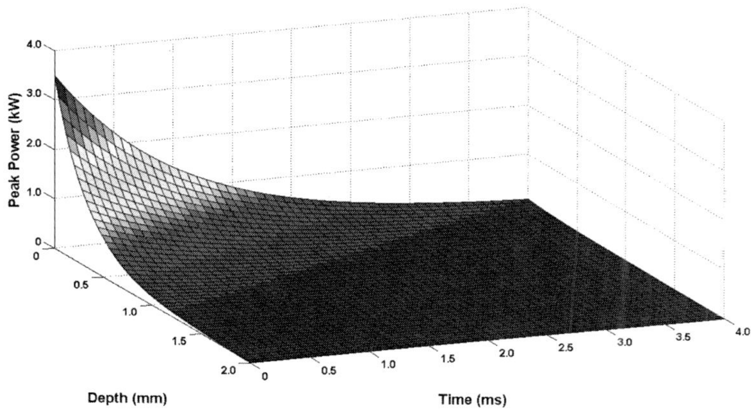

FIGURE 6. Peak power versus depth and time for stainless steel with absorption coefficient, =0.3 at the centre of penetration.

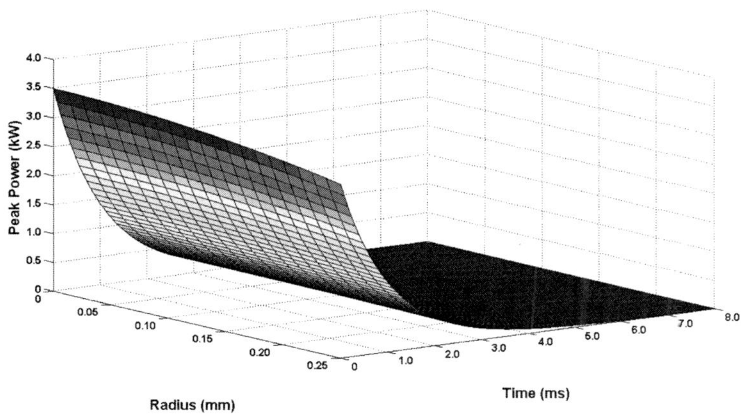

FIGURE 7. Peak power versus radius and time for stainless steel with absorption coefficient, =0.3 at the welding front.

4. ANALYSIS OF THE PULSED ND:YAG LASER SPOT WELD

The profiles for the penetration depth and bead width produced by the pulsed Nd:YAG laser beam are depicted in Figure 8 and Figure 9. The weld dimensions (weld penetration depth and bead width) for different laser beam peak powers are illustrated in Figure 10. The results display an increase in penetration depth and bead width with an increase in the laser beam peak power. The deepest penetration depth produced is 1.31 mm and the largest bead width is 0.57 mm when laser beam peak power is set at 3.5 kW. When laser beam peak power is reduced to 0.5 kW, the reading of penetration depth and the bead width are only 0.36 mm and 0.24 mm. The linear gradients of penetration depth and bead width are 0.301 and 0.10, respectively. These values show the laser beam peak power is almost 3 times more effective on the penetration depth rather than the bead width. This suggests that the laser beam peak power is a reliable parameter to control the desired penetration depth. It is observed in Figure 10 that the penetration depth and bead width increases when the pulse duration is increased. From the Figure 11, linear gradient of penetration depth and bead width are 0.0039 and 0.035, respectively. Only slight difference is noted and this means that the pulse duration has no significant effect either on penetration depth or bead width. As compared with the effect of laser peak power, the pulse duration is a much more better parameter to control the desired bead width.

Laser beam can be defocused by moving the focus point position forward or backward from the specimen surface. Figure 12 depicts that the penetration depth decreases significanly when the laser beam focus point position is moved from 0 to 4.0 mm with respect to the specimen surface. After 4.0 mm, the penetration depth decreases gradually untill 6.0 mm and there is no penetration depth that can be traced after 6.0 mm. Otherwise, when the focus point position is moved forward or backward, the bead width increases. But after 6.00 mm, there is no more bead width and only the sign of burning can be observed on the welding surface. As the focus point position is moved forward or backward from the specimen surface, the laser spot size increases. This will reduce the intensity of the laser beam which is given by, $I0=P/\pi r^2$. Here, P is the laser beam peak power and r is the laser spot radius. It indicates that the intensity of laser beam is negatively proportional with r^2. Hence, the laser beam does not have sufficient intensity to penetrate the material. As illustrated in Figure 13, it is observed that the penetration depth and bead width changes with the laser beam incident angle. As the laser beam incident angle increases, the laser spot becomes elliptical and wider. Hence the bead width also becomes elliptical and wider. The wider laser spot will reduce the intensity of the laser beam. This will result in a decrease of the penetration depth. However, when the laser beam incident angle is increased, the reflectivity of material surface drops due to the influence of light polarization. When the reflectivity drops, it will transport more laser energy into the welding material. Hence, it will increase the penetration depth. It can be seen that when weld penetration depth at angle of incidence of 45 degrees is much higher than that at 25 degrees. But, even though the reflectivity of the material surface drops greatly at 65 degrees, the penetration depth is only 0.82 mm because laser beam is less intense due to the significant increases in the laser spot area for a slight increase of the incident angle. InFigure 14, it can be seen that the penetration depth increases slightly with the number of shots. The penetration depth for the first and seventh pulse shcts are 0.91 mm and 1.11 mm. When the first shot is applied it produces a shallow concave hole on the specimen surface. This is due to the material ablation produced by laser pulse pressure when it strikes the material surface. Then, when the second shot is given, it is able to go towards the bottom of the concave hole. The second shot weld penetration depth is similar to the first shot but with an increase in the depth of the concave hole. The first and seventh shot produce 0.52 mm and 0.57 mm of bead width. The difference is relatively small because the same laser beam parameters are used for each shot.

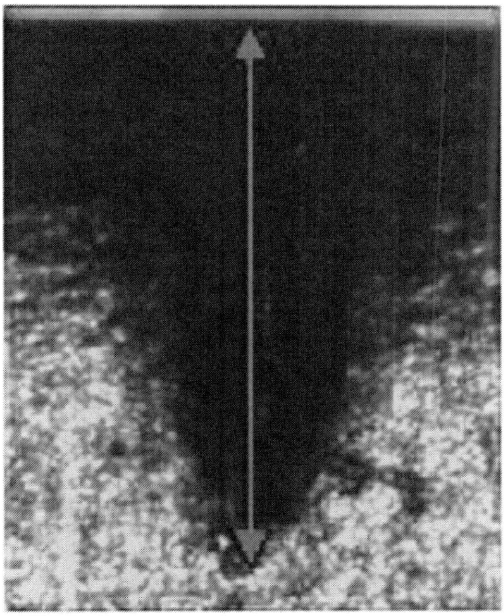

FIGURE 8. Cross-section of a spot weld produced by 2.5kW laser beam peak power and 2.5ms pulse duration. The penetration depth is labelled by the vertical line.

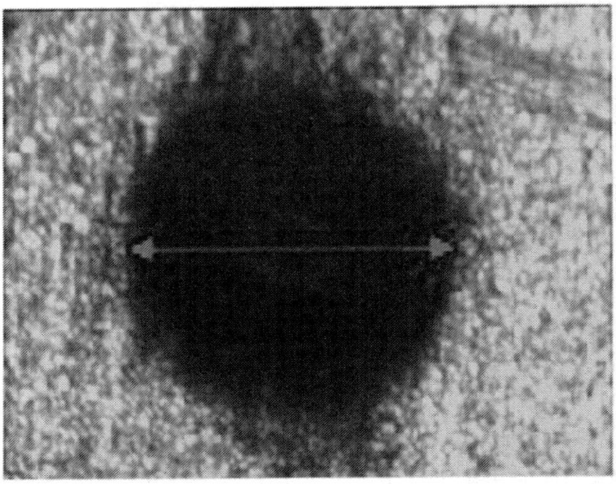

FIGURE 9. Figure 9. Top view of a spot weld produced by 2.5kW laser peak power and 2.5ms pulse duration. The bead width is labelled by the horizontal line.

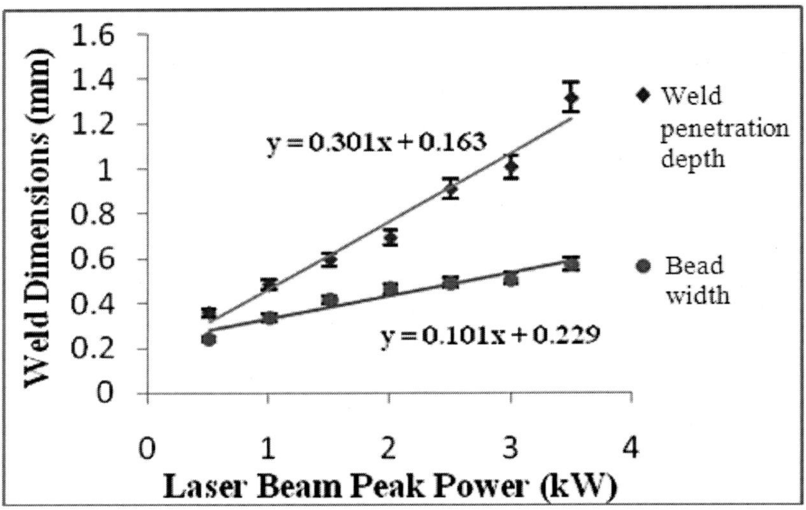

FIGURE 10. Characteristics of a spot weld dimensions as a function of laser peak powers conducted by 2.5 ms pulse duration with laser beam incidence is vertical in respect to the surface normal.

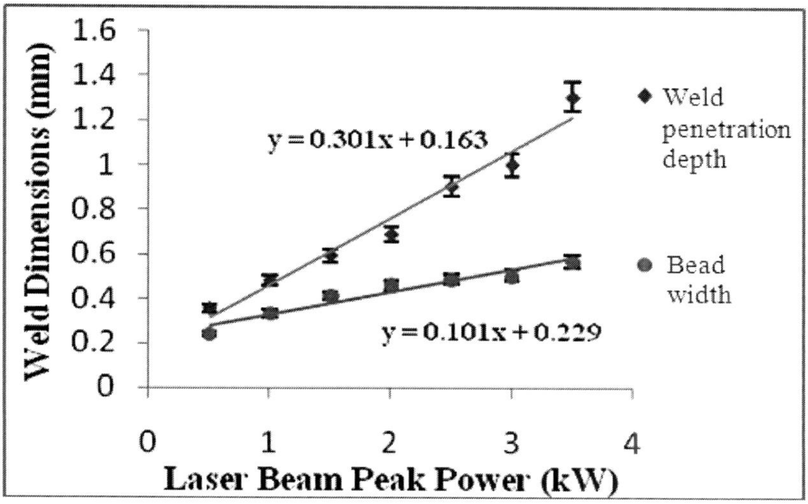

FIGURE 11. Characteristics of a spot weld dimensions as a function of laser pulse durations conducted by 2.5 kW laser peak power with wavelength of 1064 nm.

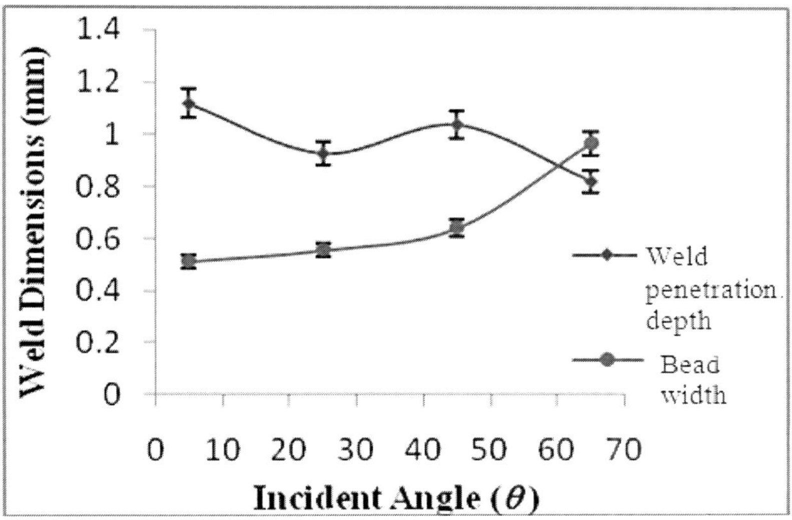

FIGURE 12. Characterictics of a spot weld dimensions when Nd:YAG laser spot is positioned forward and backward from the stainless steel 304 surface with 2.5 kW laser peak power and 2.5 ms pulse duration.

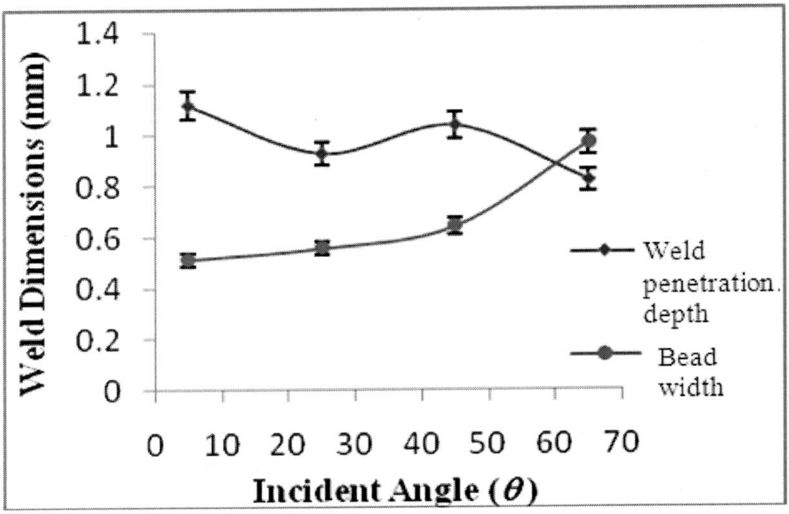

FIGURE 13. Characterictics of a spot weld dimensions when Nd:YAG laser spot incident angle is varied with respect to the stainless steel 304 surface normal by employing 2.5 kW laser peak power and 2.5 ms pulse duration.

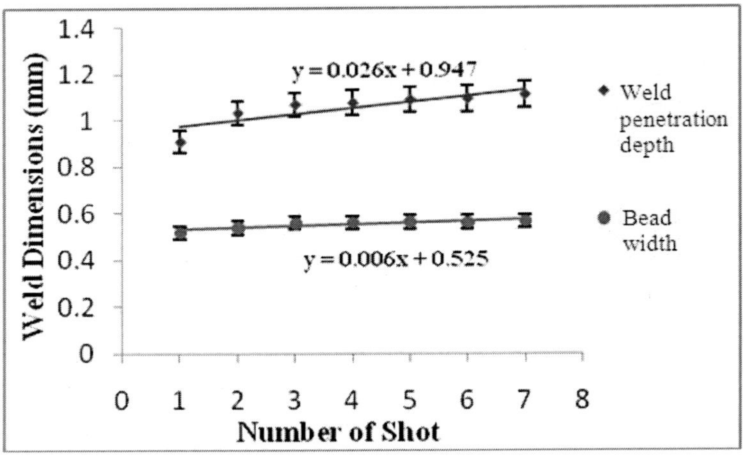

FIGURE 14. Spot weld produced by 2.5 kW laser peak power, 2.5 ms pulse duration with laser beam incidence is vertical in respect to the surface normal and laser focus point is on the stainless steel 304 surface, while the laser beam shot numbers are varied.

5. THE STRENGTH ANALYSIS OF A WELDING JOINT

In this section, the strength test of a welding joint of a single spot welding is discussed for stainless steel 304. A test is also executed for a seam welding which is produced by shooting the laser pulses continuously along the 10.0 ± 0.1 mm stainless steel 304 interfaces. The strength for seam welding of stainless steel 304 is also compared to an Invar™, which is a commercial welding material for photonics device packaging. A Unitek Miyachi LW10E ultra compact pulsed Nd:YAG laser with wavelength of 1.064 μm is employed to produce a weld joint. The energy per pulse output is in the range of 1 to 20 J, laser beam peak power is up to 3.5 kW and pulse duration ranging from 0.3 ms to 10.0 ms. However, in this investigation, only 3.5 kW laser beam peak power and 6.5 ms pulse duration is employed. This parameter is chosen because it produces a deepest penetration depth on stainless steel 304 which is 1.2 mm (Nawi et al., 2011). The penetration depth is important and should be large enough to achieve a strong attachment. The welding base material used is a stainless steel 304 sheet with a thickness of 1.0±0.1mm. The experimental setup to produce a weld consists of a laser source, fiber optics delivery system, and focusing lenses with focal length of 100.00 mm as illustrated in Figure 14.

The optical fiber cables transmit the laser beam to the welding focusing lenses inside the lens housing. The laser welding system is also equipped with an aiming diode laser beam, which simplifies positioning of the laser spot on the stainless steel sheet. A spot weld and a seam weld are produced with two types of joint; butt joint and lap joint, as illustrated in Figures 15(a) and 15(b). The strength of the weld joints is conducted by pulling the joint materials using INSTRON Series IX/s Automated Materials Tester System. The crosshead speed utilized for pulling the joint is set to 0.2 mm/min. The strength for a weld joint is determined by measuring the applied load during the test.

The element composition of stainless steel 304 and welded stainless steel 304 are examined by using Energy Dispersive X-ray (EDX) technique. The stainless steel 304 consists of iron (Fe) 69.59wt%, chromium (Cr) 18.33wt% and nickel (Ni) 12.08wt%. Figure 16 shows the element composition for stainless steel 304 changes when it is welded by Nd:YAG laser beam and it is now comprising of iron (Fe) 68.83wt%, chromium (Cr) 17.10wt% and nickel (Ni) 14.08wt%. It is observed that there are no changes in the elemental composition before and after welding of stainless steel 304. The applied load for the test and the pulling displacement for spot and seam welding are shown in Figure 17 and Figure 18, respectively. Maximum applied load is considered as the strength of the weld joint. Figure 16illustrates maximum load for butt and lap joint which are

0.2296 kN and 0.0691 kN with a pulling displacement of 0.43 mm and 0.40 mm, respectively.

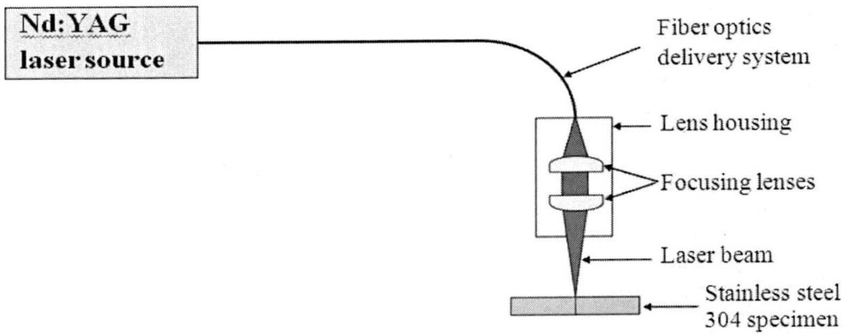

FIGURE 15. Nd:YAG laser experimental setup to produce a spot weld on stainless steel 304 specimen.

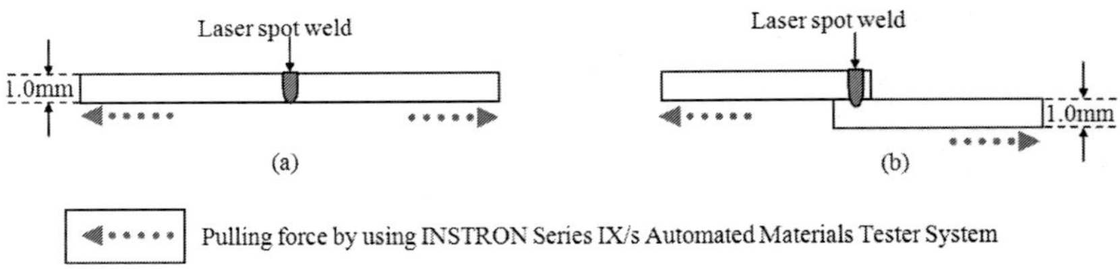

FIGURE 16. Configuration of the (a) butt joint and (b) lap joint, for stainless steel 304 with a thickness 1.0±0.1mm.

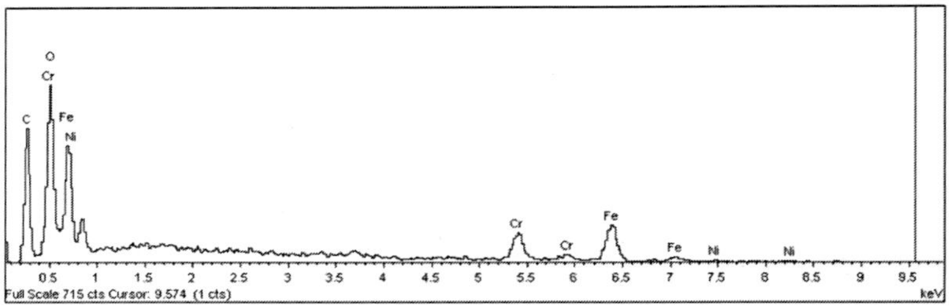

FIGURE 17. EDX analysis for the elemental composition of welded stainless steel 304.

The Invar™ maximum load is recorded as 0.3760 kN with 0.43 mm pulling displacement. The outcome indicates that the butt joint produces much stronger attachment than the lap joint. The stronger attachment is contributed by a larger volume of welded stainless steel at the joint interfaces as shown in Figure 19(a). For lap joint, the thickness of the stainless steel 304 becomes double as the upper and lower stainless steel sheets are overlapped. If the laser beam produces insufficient penetration depth, thus only a small volume of welded stainless steel 304 is produced at the interfaces of the joint as shown in Figure 19(b). This will lead to a weaker joint. But, the penetration depth can be increased by controlling the laser beam parameters such as laser beam peak power and pulse duration. Hence, it will increase the strength of a lap joint. In terms of welding materials comparison, the commercial welding material, Invar™, provides more robust joint than stainless steel 304. Figure 18 indicates the seam welding maximum load for butt joint is 2.6339 kN with 0.62 mm pulling displacement. Lap joint recorded the maximum load of 1.0466 kN at 0.49 mm. This implies that the butt joint of a seam welding provides a much stronger attachment than the lap joint similar with the case of a single spot weld. Figure 18 also indicates between the butt joint pulling displacement of 0.48 mm and 0.72 mm, there are no significant changes in the joint strength because each point along the seam welding has the similar maximum strengths but it is not achieved simultaneously depending on its geometry.

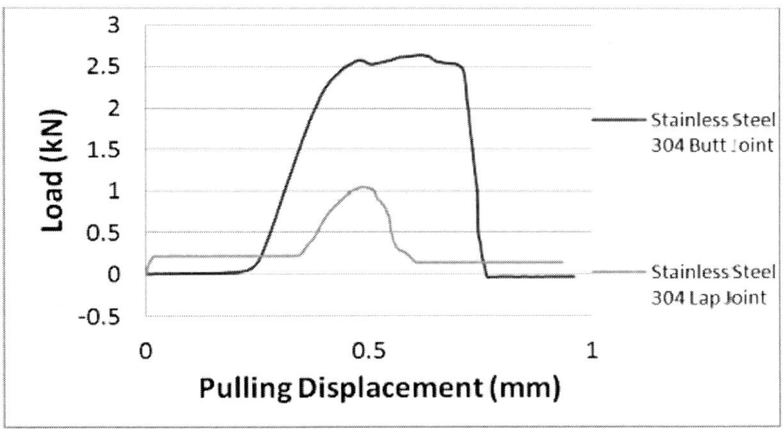

FIGURE 18. Strength profile of a single spot weld produced on stainless steel 304 and Invar™ by using 3.5 kW Nd:YAG laser beam peak power and 6.5 ms pulse duration.

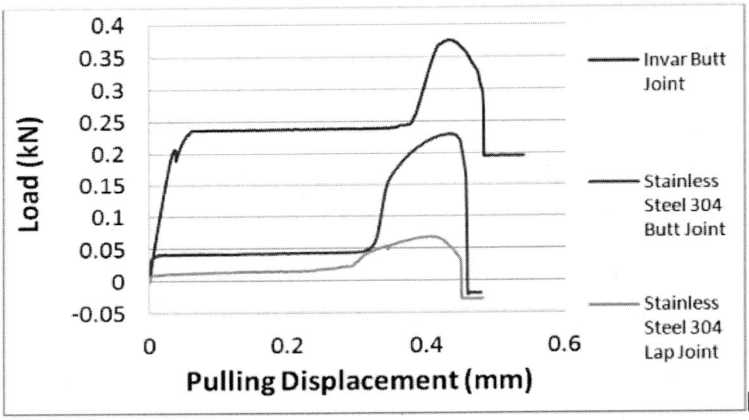

FIGURE 19. Strength profile of a seam weld produced on a stainless steel 304 along 10.0 ± 0.1 mm joint interfaces.

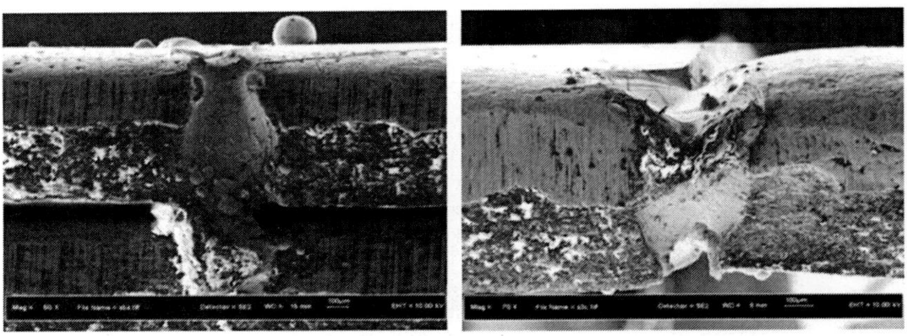

a) b)

FIGURE 20. FESEM micrographs of a stainless steel 304 spot weld joint; (a) butt joint and (b) lap joint, produced by 3.5 kW laser beam peak power and 6.5 ms laser pulse duration.

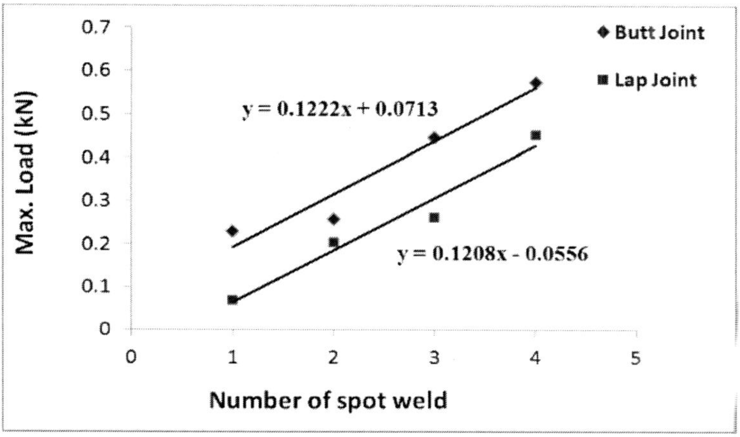

FIGURE 21. The strength comparison between butt joint and lap joint for variation number of spot weld.

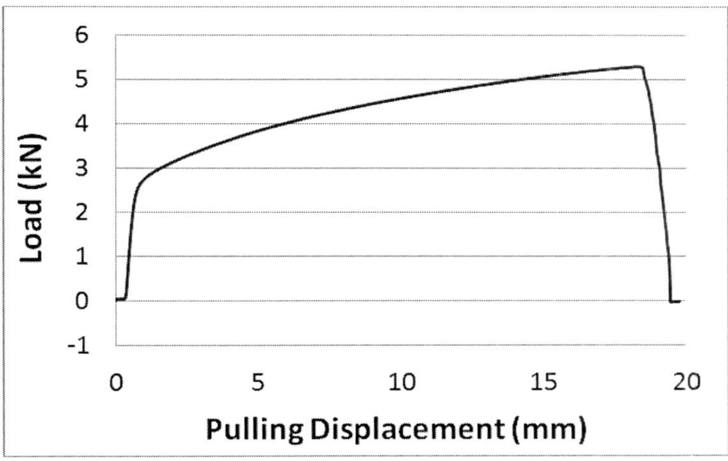

FIGURE 22. The strength of a stainless steel 304 with 1.0 ± 0.1 mm thickness and 10.0 ± 0.1 mm width.

In miniature assemblies, only a small number of laser spot weld is used for an attachment because of the relative small parts (Fadhali et al., 2007a, 2007b, 2007c, 2007d, 2008). The strength on the number of laser spot welding up to four spots is investigated for butt joint and lap joint. Then, the maximum strength for both joint types is compared. Figure 20 illustrates that the strength increases when the number of spots increases for butt and lap joint. A linear relationship is obtained from both joints strength against the number of spots and it has equal gradient which are 0.122 and 0.120, respectively. This suggests there is no significant difference in the influence of number of spots on the strength for both types of joint. The strength of stainless steel 304 is shown in Figure 21. It produces a strength of 5.2866 kN with 18.23 mm pulling displacement. The strength of stainless steel 304 is almost twice stronger than the seam welded stainless steel 304. Thus the welded material cannot attain the strength of the base material itself. The pulling displacement indicates that the elastic modulus of stainless steel 304 is much more greater than the welded stainless steel 304.

6. CONCLUSION

In conclusion, it is shown that the penetration depth and bead width increase when the laser beam peak power, pulse duration and number of shot are increased. It is found that the laser peak power is more effective to produce the deeper penetration depth rather than the bead width. Otherwise, the laser beam pulse duration is an accurate parameter to control if the desired bead width is required rather than the penetration depth. As the focus point positions are placed forward and backward from the stainless steel 304 specimen surface, the laser spot size increases. This leads to the reduction of the laser beam intensity. Hence, the penetration depth decreases. The increase of the laser spot size will increase the bead width. When the laser beam incident angle is varied, the bead width also changes due to the widening of the laser spot size. The penetration depth depends on the both widening laser spot size and material surface reflectivity that drops due to the influence of light polarization. The penetration depth increases slightly with the number of shots due to the material ablation produced by laser pulse pressure when it strikes the specimen surface. For welding joint strength, a butt joint produces much stronger attachment than a lap joint. The stronger attachment is contributed by a larger volume of welded stainless steel 304 at the joint interfaces. It is observed that each point along the seam welding has a similar maximum strengths but it is not achieved simultaneously. The similar gradient obtained in a linear relationship suggests that there is no significant difference in the influence of number of spots on the butt and lap joint strength. The commercial welding material, Invar™, provides more robust joint rather than stainless steel 304. The welded stainless steel 304 cannot attain the strength and elastic modulus of the stainless steel 304 itself. From the strength test results, it is found that stainless steel 304 is a good candidate to be used as a welding material for photonics device packaging by using pulsed Nd:YAG laser welding technique.

References

1 J. R. Beretta, W. Rossi, M. D. M. Neves, L. A. Almeida, N. D. Vieira, 2007Pulsed Nd:YAG laser welding of AISI 304 and 420 Stainless Steel. Journal of Optical Lasers Engineering, 45960966

2 C. Dawes, C. , 1992Laser welding: A practical guide. Woodhead Publication Ltd., England.

3 M. Fadhali, J. Zainal, Y. Munajat, J. Ali, R. Rahman, 2007aReliable pigtailing of photonic devices employing laser microwelding. Journal Engineering and Applied Science, 217241728

4 M. Fadhali, J. Zainal, Y. Munajat, J. Ali, R. Rahman, 2007bInvestigation of the application of Nd:YAG laser welding to couple photonic devices and packaging. Lasers Engineering, 17273286

5 M. Fadhali, J. Zainal, Y. Munajat, J. Ali, R. Rahman, 2007cLaser diode pigtailing and packaging using Nd:YAG laser welding technique. Jurnal Komunikasi Fisika Indonesia, 5203208

6 M. Fadhali, J. Zainal, Y. Munajat, J. Ali, R. Rahman, 2007dAnalysis of laser microwelding applied for photonic devices packaging. Journal Solid State Science and Technology Letter, 141819

7 M. Fadhali, 2008Efficient coupling and reliable packaging of photonic devices using laser welding technique. Ph.D Thesis: Universiti Teknologi Malaysia, Skudai, Johor, Malaysia.

8 K. Kazemi, J. A. Goldak, 2009Numerical simulation of laser full penetration welding. Computer Material Science, 44841849

9 C. Marley, 2002Laser welding photonics devices: Proper design guidelines enable users to achieve high-precision alignment requirements. Industrial Laser Solutions for Manufacturing, 2002.

10 S. A. A. Mousavi, A. R. Sufizadeh, 2008Metallurgical investigations of pulsed Nd:YAG laser welding of AISI 321 and AISI 630 stainless steels. Journal of Material Design, 154449

11 I. Naim, T. Saktioto, M. Fadhali, J. Ali, P. P. Yupapin, 2009An investigation of a pulsed Nd:YAG laser welding technique. Proceedings of IEEE-Regional Symposium on Micro and Nanoelectronics, 274280Kelantan, Malaysia, August 10-12, 2009.

12 I. Naim, T. Saktioto, M. Hamdi, M. Fadhali, J. Ali, P. P. Yupapin, 2010Penetration depth estimation for stainless steel 304L using Nd:YAG laser spot micro welding. Proceedings of 2nd Topical Meeting on Lasers and Optoelectronics, 1278288Terengganu, Malaysia, March 13-15, 2010.

13 I. N. Nawi, Saktioto, M. Fadhali, M. S. Hussain, J. Ali, P. P. Yupapin, 2011A Reliable Nd:YAG Laser Welded Stainless Steel 304 for Photonics Device Packaging. Procedia Engineering, 8380385

14 M. Pang, G. Yu, H. H. Wang, C. Y. Zheng, 2008Microstructure study of laser welding cast nickel-based superalloy K418. Journal of Material Processing Technology, 207271275

15 J. F. Ready, 1997Industrial applications of lasers. New York: Academic Press.

16 J. Zhou, H. L. Tsai, 2008Modeling transport phenomena in hybrid laser-MIG keyhole welding. Internattonal Journal of Heat and Mass Transfer, 5143534366

12

CHAPTER

Numerical Analysis of Two-Dimensional Welding Process Using Particle Method

Shun Saso, Masashi Mouri, Manabu Tanaka, Seiichi Koshizuka

Joining and Welding Research Institute, Osaka University, Osaka, Japan

ABSTRACT

In recent years, many structures were made by welding. On the other hand, new materials are used to improve the performance of the structure. However, the welding of a new material is difficult and it is difficult to predict the best condition of welding. It is hoped to develop a numerical analysis method to predict welding condition of a new material. The welding arc can be calculated by some grid method. But it is difficult to calculate the surface shape of welding pool by grid method because the surface shape of welding pool and weld penetration is deformed violently by some force. Moving particle semi-implicit (MPS) method is the numerical method which can be applied because the shape of surface is calculated easily. This method is good at calculation of large deformation and matches numerical analysis of welding process. In the present study, a particle simulation model is developed. The convection in welding pool was analyzed using these models. The Marangoni force is worked toward low temperature side in calculation. Then, the calculated weld penetration became wide and shallow. On the other hand, in calculation, the Marangoni force is worked toward high temperature side. The calculated penetration becomes narrow and deep.

Keywords (IIW Thesaurus) Welding Heat Particles Simulating

1 INTRODUCTION

Welding is a technology that melts a material and filler metal by the plasma arc. The surface shape and the penetration shape are deformed by the influence of four types of convections, as shown in Fig. 1. The four types of convections interact with each other. As a result, complicated flow will occur in welding pool. In welding pool, the heat energy is transported by the flow, and the shape of the welding pool changes [1]. For example, in the case of a strong outward convection, the heat energy will be transported for outside, and the weld penetration becomes shallow and wide. In contrast, in the case of a strong inward flow, the heat energy will be transported along the depth direction, and the weld penetration becomes narrow and deep. Therefore, the shape of the bead and penetration of the weld zone can significantly affect the mechanical performance of a joint, and hence, predicting the shape created by welding is essential. However, prediction via conventional computational fluid dynamics is difficult because the melting section is deformed greatly by arc affect and convections.

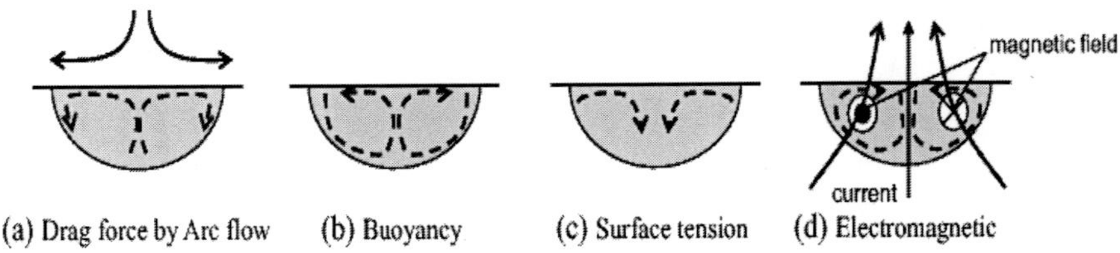

Figure. 1 Convection in a welding pool

On the other hand, the moving particle semi-implicit (MPS) method developed by Koshizuka [2–4] expresses fluid flow as a collection of particles. This facilitates the calculation of large deformations, free surfaces, and the breakup or union of fluids.

However, many other particle methods such as smoothed particle hydrodynamics (SPH) and molecular dynamics (MD) are currently available. SPH can be used for various problems, but the governing equation is difficult and computationally expensive. MD is also very useful; however, molecular behavior is determined on a nanoscale. The size of the welding area varies from millimeters to meters, and hence, MD calculations are unsuitable for the simulation of welding. If welding simulations are to be used in the production field, then simple and fast calculations are required. The MPS method is easy to use and requires less computational effort than other particle methods. Therefore, we believe that the MPS method is well suited for the analysis of welding.

In this study, we determine the forming mechanism of a welding pool. To determine welding pool phenomena, we developed an MPS method that incorporates a thermal conduction model, heat transfer and radiation model, melting and solidification model, and surface tension model. However, determining both the welding arc and welding pool phenomena using the MPS method is difficult because it is extremely time consuming and has high calculation cost. As such, welding arc analysis is performed using the lattice-based Finite Volume Method (FVM). To reduce computational cost, welding arc data are then used by the MPS method as boundary conditions or source calculations. In this study, we describe the addition of a physical model to the MPS method and report the results of the welding pool analysis.

2 MOVING PARTICLE SEMI-IMPLICIT

2.1 Moving particle semi-implicit

In this study, we use the MPS method developed by Koshizuka [2–4] to perform an incompressible flow analysis. We also use the program code provided by the "*MPS Code User Group*." The momentum equation of an incompressible fluid is given as

$$\frac{Du}{Dt} = -\frac{1}{\rho}\nabla P + \upsilon\nabla^2 u + g + F_s + F_{ma}. \tag{1}$$

Here, u, ρ, P, v, g, F_s, and F_{ma} are the velocity, density, pressure, kinematic viscosity, gravity, surface tension, and Marangoni force, respectively. Equation (1) includes both gradient and Laplacian differential operators. These operators are discretized by the particle interaction model of a common MPS method.

Equation (1) is discretized using a numerical model for Laplacian and gradient operators; this model considers the interaction between the particles. The gradient differential operator is given as

$$\langle\nabla\varphi\rangle_i = \frac{d}{n^0}\sum_{j\neq i}\left[\frac{\varphi_j-\varphi_i}{\left|r_j-r_i\right|^2}(r_j-r_i)w(\left|r_j-r_i\right|)\right]. \tag{2}$$

The Laplacian differential operator is given as

$$\langle \nabla^2 \varphi \rangle_i = \frac{2d}{n^0 \lambda} \sum_{i \neq i} \left[\left(\varphi_j - \varphi_i \right) w\left(\left| r_j - r_i \right| \right) \right]. \tag{3}$$

Here, φ is a physical quantity, d is the number of dimensions, n^0 is particle number density at the initial condition of a disposed particle, r is the position of the particle, and $w$is the weight function. The weight function w varies with the distance between the calculated particle and surrounding particles. The effect of the surrounding particles is considered by the weight function w when their interaction with the calculation particle is determined. Moreover, particle i refers to the calculated particle whose number density $n_i$is defined as follows [2–4]:

$$w(r) = \begin{cases} \frac{r_e}{r} - 1 & (0 \leq r \leq r_e) \\ 0 & (r_e \leq r) \end{cases}, \tag{4}$$

$$n_i = \sum_{j \neq i} w\left(\left| r_j - r_i \right| \right). \tag{5}$$

Here, r_e is the effective radius. The effect of other particles on the effective radius of particle i is considered in the calculation; r is the distance between particles i and j, and $r_i$and $r_j$are their respective positions. Furthermore, particles j are the particles surrounding particle i in a volume of effective radius r_e.

λ is a parameter used for normalization in Laplacian model and is given as follows:

$$\lambda = \frac{\sum_{j \neq i} \left| r_j - r_i \right|^2 w\left(\left| r_j - r_i \right| \right)}{\sum_{j \neq i} w\left(\left| r_j - r_i \right| \right)}. \tag{6}$$

The momentum and energy equations are determined from Eqs. (2)–(6).

A temporary velocity is calculated and used to determine the gravity and surface tension terms via an explicit calculation; the velocity is then corrected by a pressure equation. The particle number density is changed by the temporary velocity. This density is proportional to that of the material. In fact, instead of using the material density, pressure can be described in terms of the initial particle number density n^0 and temporary particle number density n^*.

$$\nabla^2 p^{k_n+1} = -\frac{\rho_0}{\Delta t^2}\frac{n^* - n^0}{n^0}. \quad (7)$$

Here, k_n is the time step of the calculation. The momentum equation is formulated as follows. First, the gravity and surface tension terms are calculated by an explicit method. Second, the viscosity term is determined by an implicit method and is used to calculate the stable high-viscosity fluid flow. Third, after moving the particle temporarily, the Poisson equation of pressure is formulated using an implicit method. However, the particle method results in unrealistic pressure oscillations at high frequencies. Therefore, we used the suppression method developed by Kondo and Koshizuka [5] to suppress these oscillations. Other calculations were performed as described in references [2–4].

2.2 **Analysis model of thermal conduction, melting, and solidification**

A law of energy conservation must be incorporated in the modeling of thermal conduction, melting, and solidification. This conservation of energy is given as follows:

$$\frac{Dh}{Dt} = k\nabla^2 T + Q, \quad (8)$$

Here, h, T, k, and Q are the enthalpy (energy per unit volume), temperature, thermal conductivity, and total amount of heat, i.e., the sum of heat input and radiation, respectively. Heat is input when $Q > 0$ and radiated when $Q < 0$. The first term on the right is discretized by the Laplacian of the MPS method, and the thermal conduction is calculated.

In this study, enthalpy at the position of a particle is calculated, and temperature is subsequently determined using Eq. (9). The temperature does not change while changing phase because it is influenced by latent heat.

$$T_i = \begin{cases} T_m + \dfrac{h-h_s}{\rho C_{Ps}} & (h < h_s) \\ T_m & (h_s < h < h_l) \\ T_m + \dfrac{h-h_l}{\rho C_{Pl}} & (h > h_l) \end{cases}. \tag{9}$$

Here, T_i is the temperature of the particle, T_m is the melting temperature, h_s is the enthalpy at the start of melting, and h_l is the enthalpy at the end of melting; C_{Ps} and C_{Pl} are the specific heat of solid and liquid, respectively. In this study, heat loss by the vaporization of the molten metal is not considered in the simulation.

The quantity of heat Q is defined as the sum of radiation q_o, heat transfer q_c, and heat input q_i.

$$Q = q_o + q_c + q_i. \tag{10}$$

The amount of radiation q_o is calculated using Eq. (11):

$$q_o = -\varepsilon\sigma\left(T^4 - T_a^4\right)S_i. \tag{11}$$

Here, ε, σ, T, and T_a are the radiation ratio, Stefan–Boltzmann constant, temperature in the heated region, and surrounding temperature, respectively. In this study, T_a is 300 K because the heat input is calculated by the arc simulation using the lattice method. If the heat input at the temperature of the atmosphere T_a is used, then the heat input will be counted twice. To avoid this double count, only heat emission is considered in Eq. (11), and the temperature of the atmosphere is considered to be 300 K.

S_i is the surface area of particle i. In the particle method, the surface area varies with the particle number density. The interparticle spacing r_0 of cubic particles is equal to the length of each side of the cube. S_i is calculated using Eq. (12):

$$S_i = \left(1 - \frac{n_i}{n^0}\right) N_s r_0^2. \tag{12}$$

Here, n^0 is the initial particle number density, n_i is the current particle number density of particle i, and N_s is the number of surface areas. For example, N_s is 4 for two-dimensional (2D) calculation and 6 for three-dimensional calculation. If n_i is 0 when there is nothing around particle i, then the surface area is $N_s r_0^2$. If n_i is n^0, corresponding to the density of the material, then the surface area is 0. Particles placed on the corner of the cube have surface areas of $0–N_s r_0^2$, which are determined via interpolation.

In addition, convection heat transfer q_c is calculated from heat transfer coefficient α and Newton's law of cooling.

$$q_c = \alpha(T_m - T_a). \tag{13}$$

The solid rate γ is defined as the rate of solid and liquid in particles and is calculated from the enthalpy [2] using Eq. (14) to calculate melting and solidification and to determine the phase of a material.

$$\gamma = \begin{cases} 1 & (h < h_s) \\ \dfrac{h_s - h}{h_l - h_s} & (h_s < h < h_l) \\ 0 & (h > h_l) \end{cases}. \tag{14}$$

Therefore, the material is completely solid when γ is 1 and liquid when γ is 0. If γ is neither 0 nor 1, the material is a mixture of solid and liquid. Moreover, reciprocal average is used for determining the thermal conductivity k of this mixed state because generally thermal conductivity k is different from liquid and solid [6].

$$\frac{1}{k} = \frac{\gamma}{k_s} + \frac{1-\gamma}{k_l}. \tag{15}$$

Here, s and l indices denote solid and liquid, respectively.

The kinematic viscosity, thermal conductivity, and specific heat are determined from table of temperature values. The values between each temperature are interpolated and used for the calculation of the particle.

Thermal conduction, melting, and solidification are calculated after being added to the computational algorithm of the MPS code. In this study, the temperature, melting, and solidification are calculated after solving the momentum equation and determining the motion of the particles.

2.3 Surface tension and Marangoni force

The improved potential model proposed by Ito et al. [7] is used to calculate the surface tension. The equation given below is used to calculate potential applied between particles.

$$\Phi(r) = C\varphi(r), \tag{16}$$

$$\varphi(r) = \begin{cases} -\frac{1}{3}\left(r - \frac{3}{2}r_0 + \frac{1}{2}r_e\right)(r - r_e)^2 & (r < r_e) \\ 0 & (r \geq r_e) \end{cases}. \tag{17}$$

Here, C is the coefficient of the potential function, and r_0 is the initial distance between the particles. The potential force F_i^{po} exerted by particle j on the effective radius of particle i is calculated from Eq. (19).

In Eq. (16), considering relations of C and surface tension coefficient σ is necessary because the potential coefficient C is unknown. However, the surface tension coefficient $\sigma$is equivalent to the energy saved in the surface per unit area. If the energy needed to create the surface is calculated, the relation between the energy and surface tension can be derived. When we divide a particle of diameter l_0 into two surfaces, each with a surface area of $l_0{}^2$, then the energy is calculated from the equation as follows:

$$2\sigma l_0^2 = \Phi(r). \tag{18}$$

If the surface tension coefficient σ is known, then the potential coefficient C can be determined from Eqs. (16)–(18).

$$F_i^{po} = \sum_j \frac{\partial\Phi(|r_j - r_i|)}{\partial|r_j - r_i|}\frac{r_j - r_i}{|r_j - r_i|}. \tag{19}$$

The potential force results in an attraction between the particles. However, forces of repulsion are also exerted from inside the fluid. The surface tension is modeled as the total force of attraction between the particles and repulsion from inside the fluid and is defined as follows:

$$F_i = \delta_s \frac{\cos\theta_i}{\sin^3\theta_i} F_i^{po}. \tag{20}$$

Here, θ_i is an angle between particle i and other particle, as shown Fig. 2. δ_s is constant. It is assumed that the surface tension affects only the outermost surface particles. However, particles that are designated as surface particles are frequently located at depths of two or three times that of the particle radius, away from the surface. To affect only the outermost surface particles, a correction coefficient δ_s is used. In this study, $\delta_s = 0.4$ was used for the calculations because this value led to more stable results than other values tested.

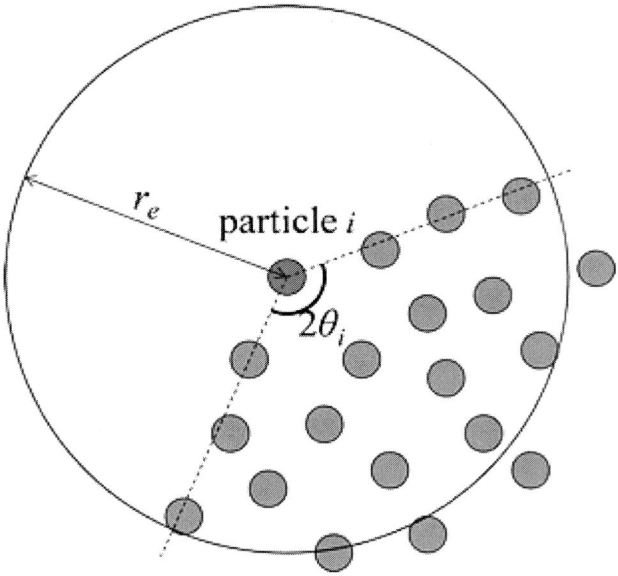

Figure. 2 Schematic showing the angle θ

We also added a wettability model such as that proposed by Ito et al. [7]; this model considers the wettability of the fluid.

The Marangoni force is a surface force that results from the temperature-dependent coefficient of the surface tension. In this study, the model developed by Koshizuka et al. [7] was used to calculate the Marangoni force using Eq. (21). Thus, the surface tension coefficient is proportional to the potential coefficient C, and the Marangoni force is determined as changes in the latter with the former, as defined in Eq. (22).

$$F_i^{ma} = \delta_m \sum_j (C_j - C_i) \frac{\partial \Phi(|r_j - r_i|)}{\partial |r_j - r_i|} \frac{r_j - r_i}{|r_j - r_i|}, \tag{21}$$

$$C_i = \frac{\sigma_i(T_i)}{\sigma_0(T_0)} C_0. \tag{22}$$

Here, C_i and C_j are the potential coefficients of particles at the fluid surface, and C_0 is the standard potential coefficient calculated from the surface tension coefficient $\sigma_0(T_0)$ at temperature T_0. As Eq. (22) shows, C_i depends on the temperature of each particle, standard potential coefficient C_0, surface tension coefficient at room temperature $\sigma_0(T_0)$, and surface tension coefficient at each particle temperature $\sigma_i(T_i)$. δ_m is constant, and $\delta_m = 0.4$ is was used for the calculations because this value led to more stable results than other values tested.

2.4 Arc force

The effect of the arc force calculated by the FVM using Tanaka's arc model [1] is considered in the MPS method as a boundary condition. In the arc calculation, two types of boundary conditions are applied at the surface of the base material, i.e., if the base material is solid, then its surface is considered as a wall, and the velocity of the solid area is zero. However, if the base material is liquid, then its surface is considered as fluid, which has the same density and viscosity as the base metal.

The node position and arc conditions (velocity and pressure) at each node calculated by the FVM are used as the background lattice in the particle method calculation. Figure 3shows a schematic of the particle positions and background lattice. In Fig. 3, the red, yellow, and gray circles represent surface, liquid, and solid particles that are affected by the arc flow, move freely, and immobile, respectively.

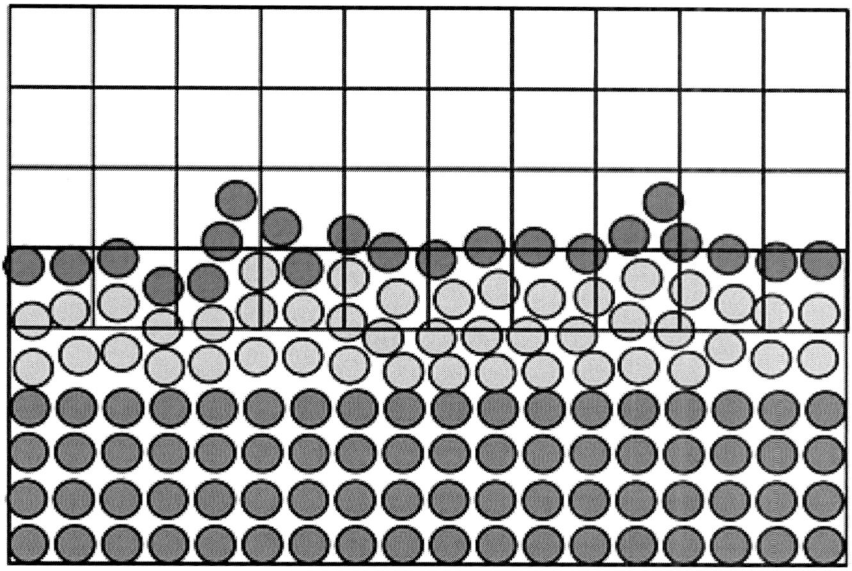

Figure. 3 Schematic of the position of the background lattice

The pressure was calculated by the MPS method using the Dirichlet boundary condition in which a pressure of zero was assigned to the outermost surface particle. However, the calculated arc pressure was applied to the outermost surface particle.

The velocity of the fluid flow is then fixed in space; because this calculation incorporates one-way coupling, the calculated velocity does not consider the shape of the welding pool surface. Therefore, the effect of the arc is reduced when the surface of the welding pool moves large distances from the background lattice. This reduced arc effect results in a rise in the surface of the welding pool. However, if the surface of the molten pool rises and the arc effects are increased, then the surface of the welding pool moves toward the arc again. This results in an instability in the welding pool behavior, and the calculation becomes unstable.

To avoid this instability, a background lattice is superimposed on the initial surface. This background lattice affects particles near the surface. In addition, changes in the shape of the surface are considered during the calculation.

As Fig. 4 shows, the flow velocity resulting from the welding arc and pressure is interpolated for particles at the surface of the background lattice because the arc force acts only on the surface of the welding pool. The velocity is calculated using the equations as follows.

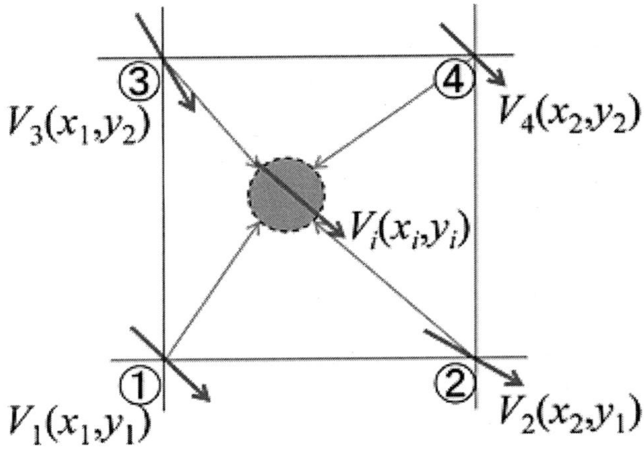

Figure. 4 Schematic of the interpolation between particles

$$\begin{cases} V_a = (V_2 - V_1)\dfrac{x_i - x_1}{x_2 - x_1} + V_1 \\ V_b = (V_4 - V_3)\dfrac{x_i - x_1}{x_2 - x_1} + V_3 \end{cases}, \tag{23}$$

$$V_i = (V_b - V_a)\frac{y_i - y_1}{y_2 - y_1} + V_a. \tag{24}$$

Here, $x_{1\text{–}2}$ and $y_{1\text{–}2}$ are the positions of nodes 1–4 in Fig. 4; x_i and y_i are the positions of particle *i*; $V_{1\text{–}4}$ is the value at nodes 1–4; V_a and V_b are the interpolation values along the*x* axis; and V_i is the value resulting from the interpolation of particle *i*. The pressure term is usually calculated using the Dirichlet boundary condition in which a pressure of zero is assigned to the surface particles. The arc pressure is the only input pressure used in this study.

The shear force at the surface of the welding pool results from the viscosity between the arc flow and surface of the welding pool. Moreover, the arc flow induces a drag on the pool. In a 2D calculation, the velocity u^*_{liquid} resulting from the arc force is determined from Eq. (25); u_{arc} and u^k_{liquid} are the fluid velocity resulting from the welding arc and convection velocity of the welding pool, respectively.

$$u_{\text{liquid}}^{*} = \frac{\rho_{\text{gas}}}{\rho_{\text{liquid}}} \nu \frac{\left(u_{\text{arc}} - u_{\text{liquid}}^{k}\right)}{r_0} \Delta t. \tag{25}$$

The density ratio $\rho_{\text{gas}}/\rho_{\text{liquid}}$ is used to account for the fact that gas and liquid have differing densities; ρ_{gas} and ρ_{liquid} are the respective densities of the arc plasma and molten metal, ν is the kinematic viscosity of the plasma, r_0 is the initial distance between the particles, and Δt is the time increment. Equation (25) is calculated explicitly before solving the Poisson equation of pressure, and u^{*}_{liquid} is defined as a virtual velocity of the particles.

3 VALIDATION OF THERMAL MODEL

3.1 Thermal conduction

The thermal calculation inserted in the MPS method was validated. One-dimensional (1D) unsteady thermal analysis was performed at a plate to validate the thermal analysis performed via the MPS method. Equation (26) was used to determine the 1D unsteady thermal conduction and was formulated to compare the results thereof with those obtained from the MPS method. Figure 5 shows a schematic illustration of the system used in this calculation. The temperatures of the iron plate and side surface were set to 300 and 500 K, respectively. The upper and bottom sides were set as the insulation boundary. In addition, the 1D temperature distribution at the center of the plate obtained via calculation from Eq. (26) was compared with the analytical solution.

$$\frac{T}{T_0} = \frac{4}{\pi}\left\{\exp\left(-\pi^2 \frac{at}{L^2}\right)\sin\frac{\pi x}{L}\frac{1}{3} + \exp\left(-(3\pi)^2 \frac{at}{L^2}\right)\sin\frac{3\pi x}{L} + \cdots\right\}. \tag{26}$$

Here, T, T_0, a, π, t, L, and x are the temperature, initial temperature, thermal diffusivity, circular constant, time, thickness of iron plate, and position of calculation point. The material-temperature-dependent shift in thermal properties was not considered in this calculation. Table 1 lists the properties of the iron plate.

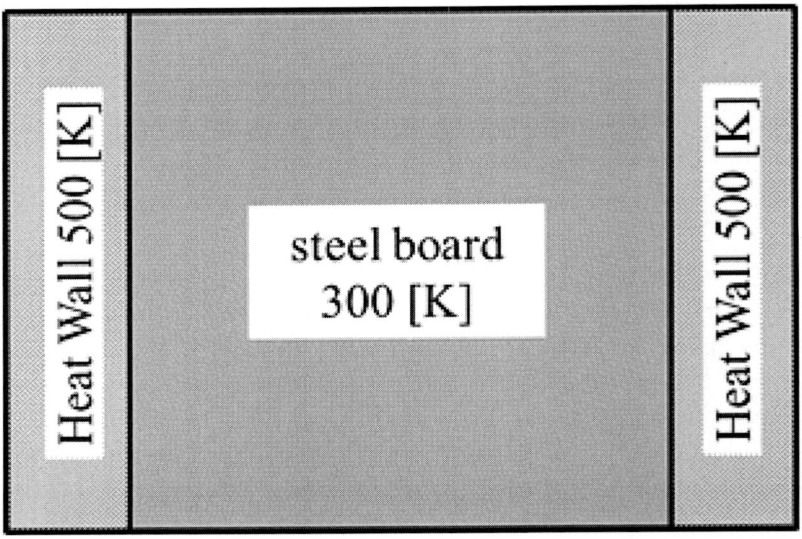

Figure. 5 Schematic of the region considered in the calculation. The *hatched and central regions*denote the high-temperature side surface and low-temperature iron plate, respectively

Table 1 Properties of the iron plate used in the thermal conduction calculation

Density ρ in kg/m^3	7870
Thermal conductivity k in W/m/K	76
Specific heat C_p in J/kg/K	460.0
Material particle number	2601

The temperature distributions obtained after $t = 300$ s were compared. Figure 6 shows the temperature distributions and corresponding graph of the analytically determined solution and the result of the MPS method. The maximum deviation between the results was 1.5 %. Furthermore, the thermal conduction was accurately modeled by the MPS method.

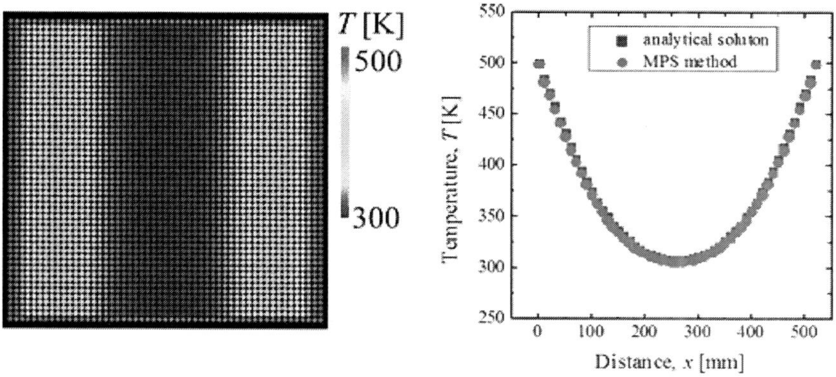

Figure. 6 The result of the MPS calculation and comparison of the 1D temperature distribution. **a**The MPS results are color coded according to temperature; **b** 1D temperature distribution obtained after 300 s. The *red and blue lines* denote the MPS and analytically determined results, respectively

Table 2 Parameters used in the calculation of heat transfer and radiation

Density ρ in kg/m^3	7920.0
Thermal conductivity k in W/m/K	26.8
Specific heat C_p in J/kg/K	641.0
Emissivity	0.9
Outside temperature in °C	25
Material particle number	25,536

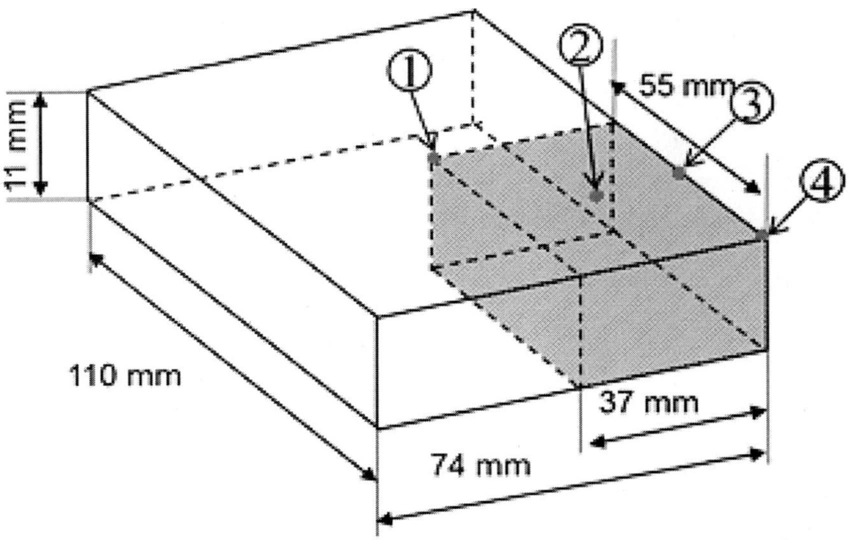

Figreu. 7 Schematic of the calculation model. The size of the calculation model was 1/4th that of 110 mm × 74 mm × 11 mm (i.e., the hatched area)

3.2 **Heat transfer and radiation**

Heat transfer was validated by investigating the cooling process of a stainless steel plate. Table 2 lists the material properties of the stainless steel used for the calculation performed under the assumptions of air atmosphere. The coefficient of heat transfer is calculated by considering natural convection. Figure 7 shows a schematic of the calculation region. Temperature was also calculated using ABAQUS and compared at points 1–4 in Fig. 4.

Figure 8 shows the result of the investigation of the cooling process at each point for 60 s. The temperatures calculated for the 60 s were compared, and a maximum deviation of 5 % was obtained between the MPS and ABAQUS results. Furthermore, deviations of 2 and 5 % were obtained in the case of points 1 and 2 that were placed at the center and edge of the top surface, respectively. The MPS result revealed a tendency for faster cooling than ABAQUS. The same element size was used in both calculations. However, the respective surface areas of the ABAQUS and MPS models were calculated as that corresponding to a square and the particle number density. Thus, the surface in the former calculation was slightly larger than that of the latter.

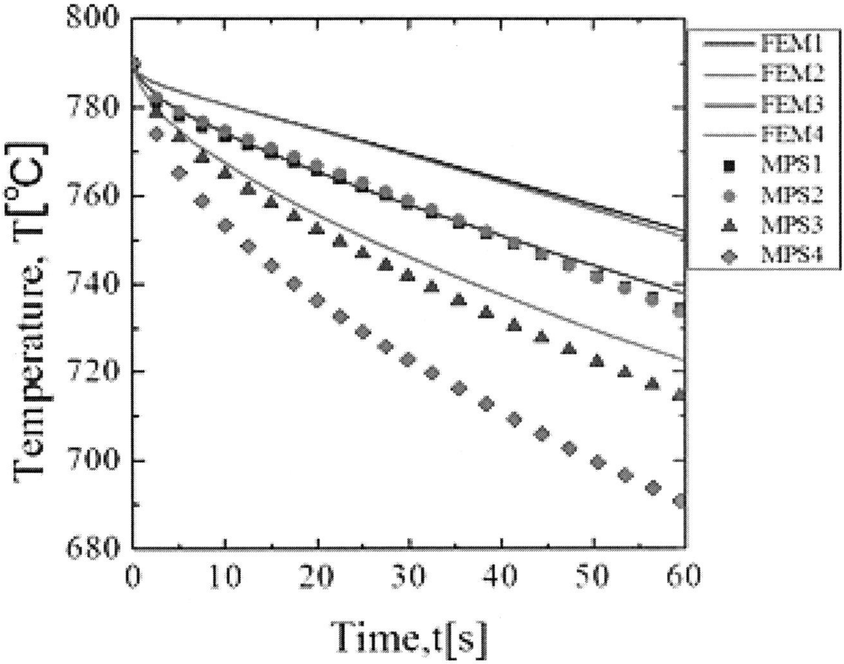

Figure. 8 Result of the cooling process of stainless steel showing temperature plotted as a function of time. Results from the ABAQUS and MPS calculations are plotted as *solid lines* and*symbols*, respectively

3.3 **Convection of buoyancy**

Buoyancy results from a temperature difference, and natural convection occurs in a fluid. A thermal cavity was determined to verify the convection of buoyancy, and the results were compared with those obtained using the FVM described by Davis [8]. Figure 9 shows a schematic illustration of the thermal cavity. The Rayleigh number (Ra) was changed by varying the temperature of the high-temperature side; the temperature of the low-temperature side was set to zero. Ras of 10^3, 10^4, and 10^5 were used in the calculation; Table 3 lists the properties of the thermal cavity considered.

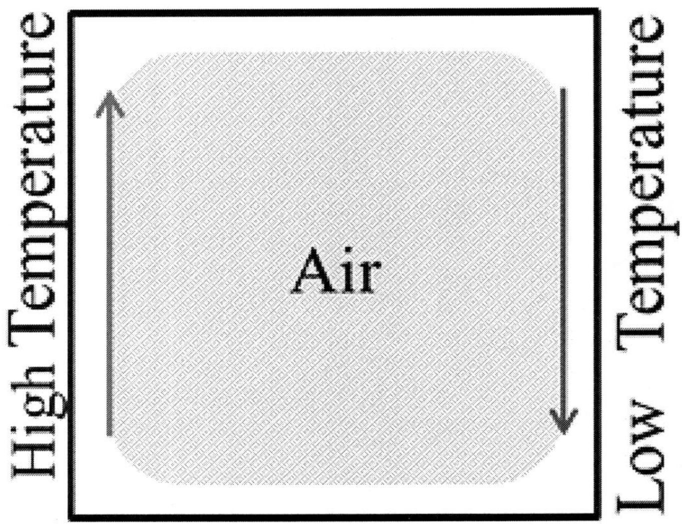

Figure. 9 Schematic of the thermal cavity. The area surrounded by the wall. One side of the wall is fixed at a high temperature and the other at 0 K

Table 3 Properties of the thermal cavity used in the calculation

Density ρ in kg/m^3	1.176
Thermal conductivity k in W/m/K	26.3×10^{-3}
Specific heat C_p in J/kg/K	1007
Expansion coefficient	3.75×10^{-3}
Material particle number	2601
Rayleigh number Ra	10^3–10^5

Figure 10 shows one of the results obtained for Ra = 10^3 and reveals a close (qualitative) correspondence between the flow line and temperature distribution of the MPS method and FVM. To compare the results quantitatively, the maximum velocity associated with each Ra result was converted to a dimensional-less variable. With this non-dimensional variable, deviation (error values) of 6, 15, and 8 % were obtained between the MPS and FVM results for Ra = 10^3, 10^4, and 10^5, respectively.

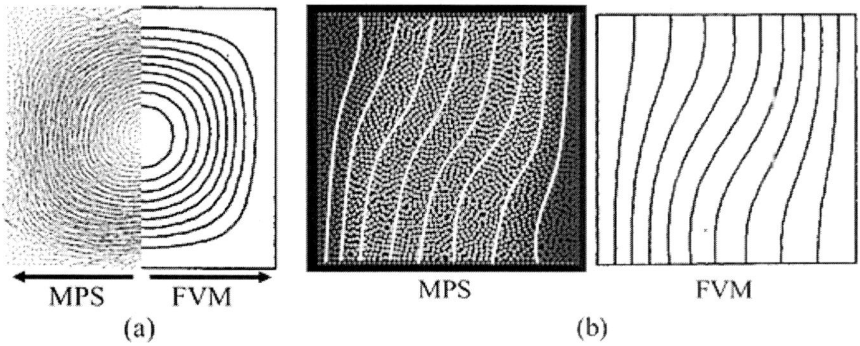

Figure. 10 Comparison of **a** flow line and **b** boundary temperature. The temperature lines were superimposed on the MPS result

3.4 **Melting and solidification**

The solidification of water was investigated to validate the implemented melting and solidification model. A 79 mm × 79 mm area was filled with water of 0 °C, which was solidified from the exterior to interior by cooling the outer wall at a rate of 0.085 °C/min. Table 4 lists the parameters used in the calculation. This problem was examined experimentally by Saitho [9]. The present calculation was based on the assumption that water flow and heat transfer did not occur. The temperature of the water was set to zero, and the thermal conductivity of ice was used for the calculation because thermal conduction occurred only between the ice and water.

Table 4 Parameters used in the solidification calculation

Phase	Water	Ice
Density ρ in kg/m^3	1000	1000
Thermal conductivity k in W/m/K	2.2	
Specific heat C_p in J/kg/K	4200	2000
Latent heat Q_l in J/m^3	3.335×10^8	
Material particle number	6400	

Figure 11 reveals a close correspondence between the calculated and experimental results for the center of each side of the cavity. However, the boundary shape at the corner was round in the case of the latter than the former. This indicates that the boundary in the MPS method had a lower cooling ratio than that in the experiment. According to Saito et al., the temperature decreases readily at the corner of the region. However, the tendency for temperature decrease was not considered in this calculation.

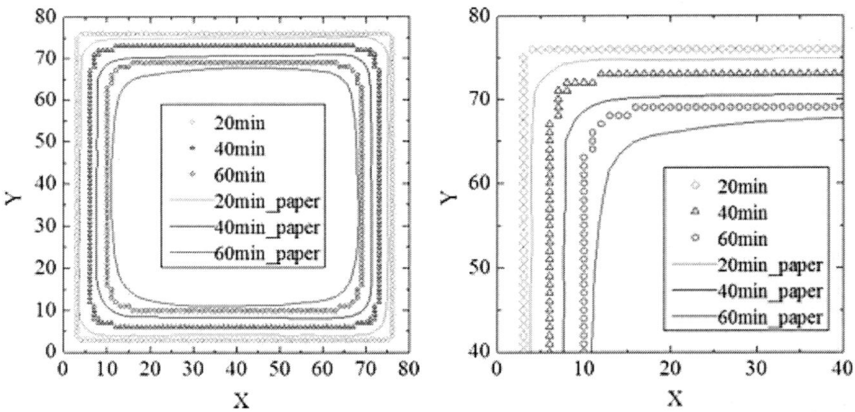

Figure. 11 The results of solidification showing the boundary between the water and ice after 20, 40, and 60 min. The *symbols* and *solid lines* denote the results of the MPS calculation and experiment by Saitho, respectively, for **a** the entire region and **b** part of the corner

4 ANALYSIS OF ARC WELDING

4.1 Conditions

The MPS method was used to calculate the welding pool in 2D, and the arc flow was considered. Furthermore, the stable TIG arc was determined by the FVM and then by the particle method. The parameters used for the arc determination included current, Ar gas flow rate, arc length, cathode tip angle, base metal, and cathode material of 200 A, 10 L/min, 5 mm, 60°, pure iron, and tungsten, respectively. The results of the calculation were output at each node. These data were then used for the particle method calculation. The approximated heat input was proportional to the distance from the center of an arc to the surface particles. Figure 12 shows the distribution of heat input; Fig. 13 shows the 2D model of the particles from the iron plate. In addition, Table 5 lists the parameters used in the calculation, and Fig. 14 shows the temperature dependence of the surface tension. The surface tension of pure iron had a negative temperature coefficient. For comparison, we also performed the calculation for a positive temperature coefficient.

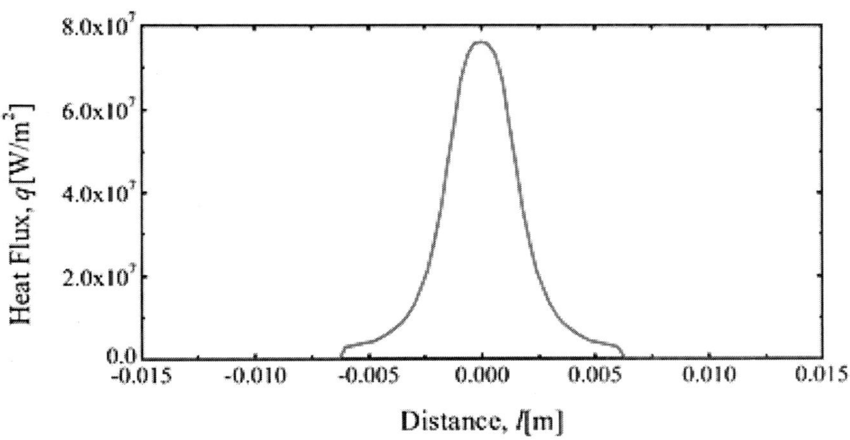

Fig. 12 Distribution of the input heat energy

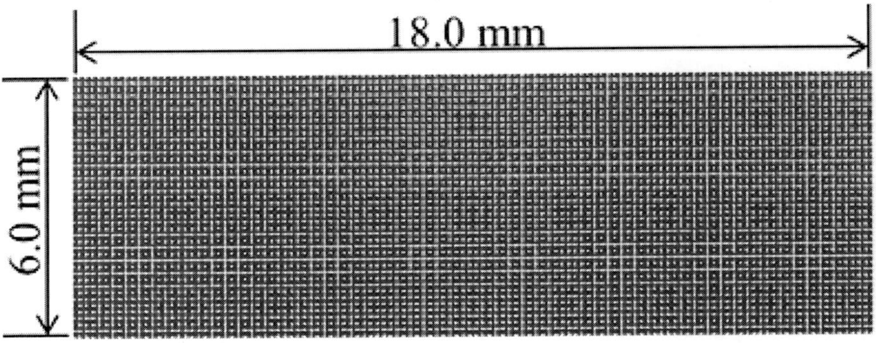

Fig. 13 Schematic of the calculation model of the particle method

Table 5 Parameters used in the weld pool calculation

Average distance r_0 in m	0.00015	
Number of particle	4961	
Time step in s	5.0×10^{-6}	
Phase	Molten iron	Solid iron
Density ρ in kg/m³	7000	7000
Thermal conductivity k in W/m/K	40.9–45.2	34.6–80.0
Specific heat C_p in J/kg/K	795	450
Melting point T_m in K	1809	
Latent heat Q_l in J/m³	1.944×10^9	
Surface tension coefficient σ in N/m	$\pm 0.49 \times 10^{-3} T + 2.625$	
Calculation time t in s	1.0	

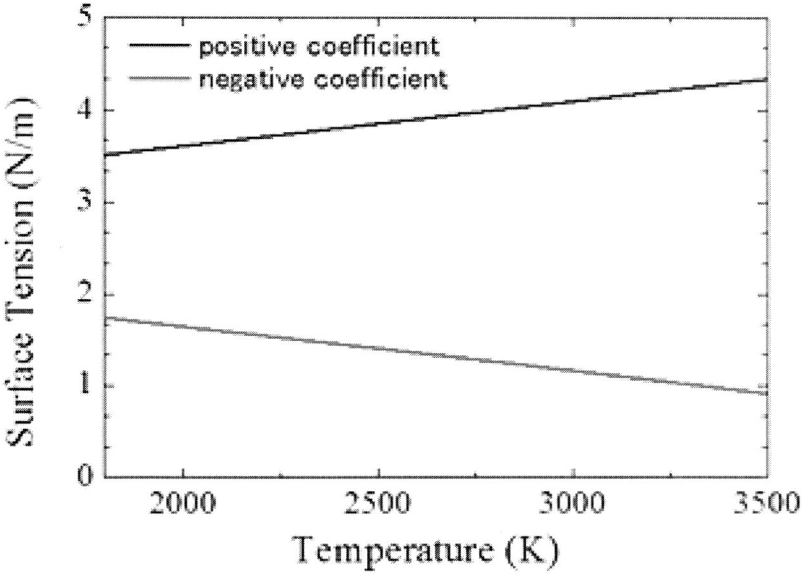

Fig. 14 Temperature dependence of the surface tension

5 RESULTS AND DISCUSSION

Figures 15 and 16 show the calculation result and velocity vector, respectively.

As Fig. 15 shows, the surface tension acts in the direction of low temperature, and outward convection was induced with the arc flow. The weld penetration became wide and shallow with heat transfer to outside via an outward flow. However, as Fig. 15b reveals, the positive surface tension acts in the direction of high temperature, thereby inducing inward convection. The resulting heat transferred in the through-thickness direction by an inward flow, and the weld penetration became narrow and deep.

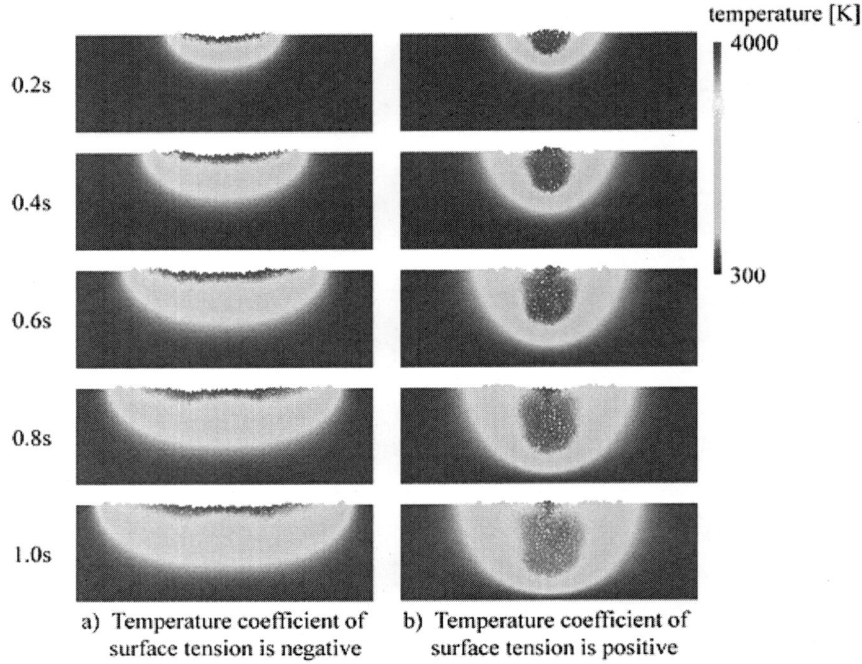

Fig. 15 Result of the welding pool analysis with the arc force color coded according to temperature for **a** negative and **b** positive temperature coefficients of surface tension

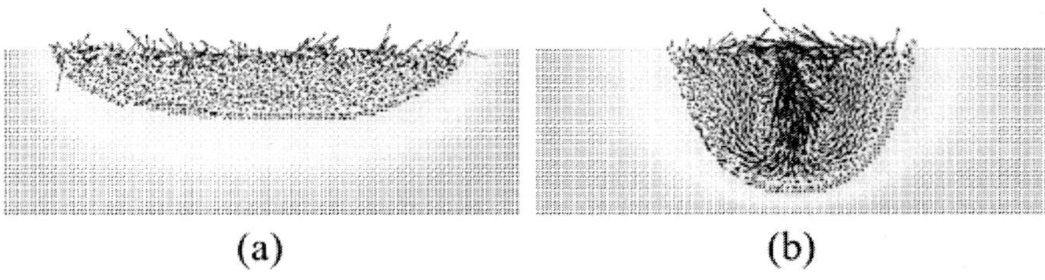

Fig. 16 Distribution of the vector for **a** negative and **b** positive temperature coefficients of the surface tension. The *arrow length* denotes the magnitude of the velocity

As Fig. 16b shows, the outward flowing arc stream is located at the surface. There is also a strong inward convection, although the velocity of the Marangoni convection is significantly smaller than the arc velocity. This stems possibly from an underestimation of the arc force defined in Eq. (25). Moreover, the influence of the arc flow should change with the deformation of the welding pool surface, and the model must be adapted or a new model must be developed to take this into consideration.

The actual material has, in fact, a negative surface tension coefficient. The result obtained using the negative coefficient was compared with the experimental result; the melting shape resulting from the MPS calculation was more than that from the experiment. This may be attributed to the thermal energy that remained in the welding pool because the input heat could not move to the front or back sides of the 3D model. Further investigation is required to explain the aforementioned difference in shape.

6 CONCLUSIONS

We developed an MPS method for modeling welding effects. This method allows the representation of a free surface, clear boundary shape, and easily calculated phase shifts.

1. To calculate the heat effect, we added a heat conduction model and a melting and solidification model to the MPS method. Surface tension and Marangoni force models were also added to determine the surface tension and Marangoni convection.
2. The effect of arc flow calculated by Tanaka's model is considered by using the one-way coupling model.
3. The calculated welding pool was shallow and exhibited wide penetration for a positive temperature coefficient of the surface tension. In contrast, the welding pool was deep and exhibited narrow penetration for a negative temperature coefficient of the surface tension. The tendency for the deformation of the welding pool was determined by the developed MPS method.

REFERENCES

1 Tanaka M, Terasaki H, Ushio M, Lowke JJ (2003) Plasma Chem Plasma Process 23:585–605

2 Koshizuka S (2005) Computational dynamics lecture series 5: particle method (in Japanese). Maruzen Company Limited

3 Koshizuka OY (1996) Moving particle semi-implicit method for fragmentation of incompressible fluid. Nucl Sci Eng 123:421–434

4 Koshizuka S, Nobe A, Oka Y (1998) Numerical analysis of breaking waves using the moving particle semi-implicit method. Int J Numer Meth Fluid 26:751–769

5 Kondo M, Koshizuka S (2008) Trans. JSCES. Paper No.20080015 (in Japanese)

6 Koshizuka S, Ohta K, Oka Y (1998) Numerical analysis of thermal-hydraulics with solidification using moving particle semi-implicit method. Proceedings of the Conference on Computational Engineering and Science 3 (in Japanese)

7 Ito J, Koshizuka S, Saso S, Mouri M (2012) Development of a particle model for marangoni force. Proceedings of the Conference on Computational Engineering and Science 17 (in Japanese)

8 de Vahl DG (1983) Natural convection of air in a square cavity a bench mark numerical solution. Int J Numer Methods Fluids 3:249–264

9 Saitho T (1976) An experimental study of the cylindrical and two-dimensional freezing of water with varying wall temperature. Technol Rep Tohoku Univ 41:61–72

INDEX